Bayes, Bugs, and Bioterrorists

Lessons Learned from the Anthrax Attacks

Kimberly M. Thompson, Robert E. Armstrong, and Donald F. Thompson

Center for Technology and National Security Policy

National Defense University

April 2005

The views expressed in this article are those of the authors and do not reflect the official policy or position of the National Defense University, the Department of Defense or the U.S. Government. All information and sources for this paper were drawn from unclassified materials.

Kimberly M. Thompson is Associate Professor of Risk Analysis and Decision Science at the Harvard School of Public Health (http://www.hsph.harvard.edu/faculty/ KimberlyThompson.html). Dr. Thompson can be contacted via e-mail at kimt@hsph.harvard.edu or by phone at (617) 432-4285.

Robert E. Armstrong is a senior research fellow in the Center for Technology and National Security Policy at the National Defense University. Dr. Armstrong may be contacted via e-mail at armstrongre@ndu.edu or by phone at (202) 685-2532.

Donald F. Thompson is a senior research fellow in the Center for Technology and National Security Policy at the National Defense University. Dr. Thompson may be contacted via e-mail at thompsond1@ndu.edu or by phone at (202) 685-2406.

Defense & Technology Papers *are published by the National Defense University Center for Technology and National Security Policy, Fort Lesley J. McNair, Washington, DC. CTNSP publications are available online at http://www.ndu.edu/ctnsp/publications.html.*

Contents

Executive Summary

The U.S. government continues to improve its plans for protecting civilians and soldiers from attacks with biological weapons. Part of this effort focuses on developing strategies that recognize the difficult choices to be made in using and deploying resources. This paper presents a risk- and decision-based framework—derived from the field of Bayesian statistics—for developing strategies that facilitate managing the risks of biological agents. The framework recognizes the significantly different attributes of potential biological weapons and offers a strategy for improving communication to effectively coordinate national biopreparedness efforts. The framework identifies generic decisions related to routine immunization, response planning, stockpiling vaccines and therapeutic agents, surveillance choices, containment, emergency response training, research, media and communications preparations, information management, and policy development. This paper provides a straw man to be used in wargames, exercises, practices, etc., at all levels of government.

Given the attention on anthrax following the 2001 attacks, this paper applies the framework to managing the risks of anthrax to provide an illustrative example. The example demonstrates that by organizing information at this level, decisionmakers can quickly understand the critical connections between different options (e.g., vaccinating with a new vaccine requires an investment in research; research might increase the opportunities for breaches of containment). With respect to managing the risks of an attack with anthrax, this analysis suggests the need for creation of a comprehensive national management plan that includes quantitative evaluation of resource investments.

The authors conclude that the government should adopt a process—based on decision science and using the power of decision trees as an analytical tool—to develop a strategy for managing the risks of bioterrorism. Using this type of approach, the government can better characterize the costs, risks and benefits of different policy options and ensure the integration of policy development. Additionally, confirmed use and refinement of decision trees during exercises will provide analysis of the long-term consequences of decisions made during an event and give policymakers insights to improve initial decisions.

Introduction

Following the anthrax attacks in the fall of 2001, the U.S. government significantly increased its commitment of resources and its efforts for biopreparedness. Some observers of the anthrax attacks may downplay the event, given the relatively small burden of illness and death that resulted (22 cases, 5 deaths)[1] compared to the potential impact from alternative methods of dispersal of weaponized anthrax. However, the U.S. government incurred significant costs associated with the attacks, which effectively created terror. Congress provided over $1 billion dollars to the U.S. Postal Service (USPS) alone for its initial interventions (including gloves and masks for workers, irradiation of mail, decontamination of facilities, and introduction of technology to protect workers against dust from letters passing through handling equipment). Cleanup of the Hart Senate Office Building exceeded $24 million, and direct cost to the USPS may exceed $3 billion.[2]

A study conducted in late 2001 suggested that as many as 2 million Americans might have taken Cipro unnecessarily,[3] which explains observed shortages in supply.[4] (See the "Too Much of a Good Thing" text box on page 22.) At a cost of approximately $600 for a 60-day supply of Cipro (approximately $60 for doxycycline), this implies additional expenditures in excess of $1 billion. Other non-traditional costs also arose in the context of the potential challenge to the patent for Cipro.[5] Collectively, this represents a fraction of the over $500 billion annual budget for the Department of Health and

---

[1] Daniel B. Jernigan et. al., "Investigation of bioterrorism-related anthrax, United States, 2001: Epidemiologic Findings." *Emerging Infectious Diseases* 8, no. 10 (October 2002): 1019.

[2] David Heyman, *Lessons from the Anthrax Attacks: Implications for U.S. Bioterrorism Preparedness, A Report on a National Forum on Biodefense* (Washington, DC: Center for Strategic and International Studies, April 2002): vii. Available at <http://www fas.org/irp/threat/cbw/dtra02.pdf>.

[3] Robert J. Blendon et al., *Harvard School of Public Health/Robert Wood Johnson Foundation Survey Project on Americans' Response to Biological Terrorism Tabulation Report* (Media, PA: International Communications Research, Oct. 24-28, 2001): 5.

[4] British Broadcasting Company, "Cipro demand outstrips supply," October 21, 2002. Available at <http://news.bbc.co.uk/1/hi/world/americas/1618783.stm>. "In New York, demand for Cipro has increased 203% from the week ending 14 September through the week ending 12 October, according to NDCHealth, a drug research company. Nationally during the same period of time, the number of Cipro pills dispensed has increased 41%, according to the company. Mr Weinstein said: 'Cipro seems to be the one that's got the most press, and people seem to think that it is the only one they think will work against anthrax.'"

[5] British Broadcasting Company, "America's anthrax patent dilemma," October 23, 2001. Available at <http://news.bbc.co.uk/1/hi/business/1613410.stm>.

Human Services,[6] but it also displaced other investments and contributed to deficit spending. The existence of weaponized anthrax with demonstrated lethality and failure to apprehend the perpetrator(s) of the anthrax letters create a demand for the U.S. government to develop a strategic plan to manage risks from anthrax and to manage resource investments at this level.

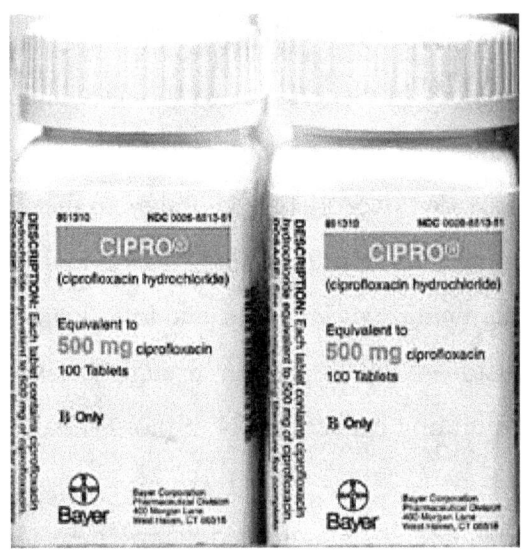

Figure 1 – Results from a 2001 survey suggest that as many as 2 million Americans may have taken the drug Cipro (ciprofloxacin hydrochloride) unnecessarily during Fall 2001—at a cost that probably exceeded $1 billion. *Image Source: David McNew/Getty Images, http://pubs.acs.org/cen/topstory/7945/7945notw3.html.*

The U.S. government continues to organize its efforts for biopreparedness and biodefense to promote homeland security. Among the government departments and agencies that play critical roles in the management of risks of biological agents are the Department of Homeland Security (DHS), charged with providing security, the Department of Defense (DOD), charged with protecting soldiers from biological weapon attacks and defending U.S. interests from attacks, the Centers for Disease Control and Prevention (CDC), the part of the Department of Health and Human Services (HHS) charged with protecting the public health, and state and local law enforcement authorities and health departments. While information sharing and planning efforts continue to evolve and improve, the nation remarkably still lacks a national plan for managing the

---

[6] United States Department of Health and Human Services, *FY2004 Budget in Brief*, 2003. Available at <http://www.os.HHS.gov/budget/04budget/fy2004bib.pdf>.

risks for even a single biological agent, and remains far from a national health surveillance and biodefense system.[7]

Coordination is a major challenge. As the response to the anthrax attacks in the fall of 2001 showed, communication and coordination make a big difference in ensuring alignment of objectives across agencies.[8] For example, the CDC and the DOD did not share the same information about the dispersal ability of the weaponized anthrax used in the attacks, and this may have contributed to confusion about the threat posed by the anthrax letters and the need for active response.[9]

Part of the challenge arises from the fact that managing the risks requires making several critical decisions related to vaccination (whom to vaccinate, which vaccine to administer), surveillance, response, post-exposure treatment (adequacy of stockpile), lab containment, and research. Developing strategies for action requires recognition that difficult choices must be made to best utilize and deploy resources and that cost-effectiveness information should guide resource decisions.[10]

The authors recognize that more work is needed to provide insight into the issues of intra- and interagency coordination and to translate theory into practice. It presents a risk- and decision-based framework for developing strategies to manage the risks of an attack with a biological agent and provides an example applied to anthrax. The paper aims to begin the work toward integration of existing efforts and improved decisionmaking and coordination. Create an ongoing process for management and to implementing strategies to better focus resources will require more coordinated effort than currently exists and a focus on better characterization of the costs, risks, and benefits of different policy options.

This paper outlines a generalized analytical approach that uses decision trees to present options and demonstrate the connections between them. (The approach is derived from the work of 18th century mathematician and theologian Thomas Bayes. See "Risk Analysis and Decision Trees," p. 9.) With this approach, analysts can quickly

---

[7] Darryl Greenwood, "Health Surveillance and Biodefense System (HSBS) Study: HSBS Feasibility and Implementation Study." *MIT Lincoln Laboratory Special Report* Volume 1 (July 2002): 90-1028.
[8] Ibid., 1026.
[9] Richard Preston, *The Demon in the Freezer* (New York, NY: Random House, 2002): 1-26.
[10] Richard Danzig, *Catastrophic Bioterrorism – What is to be Done?* (Washington, DC: Center for Technology and National Security Policy, National Defense University, August 2003): 13.

communicate with decisionmakers about the implications of combinations of options.[11]
This approach is currently being used to provide analytical support for decisionmakers
facing the complex decisions related to managing polio, once it is eradicated. In the polio
context, for example, decisionmakers currently face choices that include whether to
continue vaccination (with oral polio vaccine and/or inactivated polio vaccine), whether
to create a vaccine stockpile, and whether to continue the existing surveillance.[12] We
emphasize that this approach of focusing on decisions provides a means to cross
interdisciplinary and other boundaries (e.g., military and non-military decisionmakers,
various levels of government) and consequently it provides a useful organization and
communication tool to promote effective management.

---

[11] Nanilee K. Sangrujee, Radboud Duintjer Tebbens Victor M. Cáceres and Kimberly M. Thompson,
"Policy Decision Options During the first Five Years Following Certification of Polio Eradication."
Medscape General Medicine 5, no. 4 (December 18, 2003): 35. Available at
<http://www.medscape.com/viewarticle/464841> (subscription).
[12] Those thinking about polio only as a common infectious disease should recognize that these choices also
play a role in the risks of the potential future use of polioviruses as a biological weapon.

Thinking Big: A "Straw Man" Framework

The challenge of developing a universal framework for managing the risks of bioweapons stems in part from the reality that the potential weapons differ in important ways. For example, bioagents differ with respect to their ability to grow, spread through human contact and other vectors, disperse, change, survive in the environment, and ability to bring on lethality (in some cases the organisms pose the threat to humans while in other cases the toxins they produce pose the threat). The delayed effects of biological weapons create a window of opportunity for taking action to reduce the impact of health effects, but also allow for attacks and attackers to go undetected. The other complication associated with biological weapons comes from the reality that they often produce symptoms similar to those from natural diseases. Consequently, patients and health care providers may find it difficult to distinguish an attack from background illness.

While health care and increased vaccination continue to evolve to reduce background burdens of disease, technological advances also fundamentally change the opportunities for attackers, since they can now cultivate bioengineered microorganisms to optimize their delivery, infectivity, detectability, and treatability. The current genetic engineering techniques that automate laboratory processes give attackers more tools than ever to produce bioweapons. Of much concern to the U.S. government, some major state-funded offensive biological weapons efforts developed expertise related to bioweapons.[13]

Despite the important differences between potential biological agents, a framework can be developed that identifies the key decisions that risk managers must make for any biological agents. The authors recognize the enormous existing campaigns within different levels of the government. For example, efforts to prioritize led the Centers for Disease Control and Prevention (CDC) and the National Institute of Allergy and Infectious Diseases (NIAID) to develop and maintain a priority list of agents.[14]

---

[13] Ken Alibek, *Biohazard* (New York, NY: Random House, 1999); Judith Miller, Stephen Engelberg and William Broad, *Germs: Biological Weapons and America's Secret War* (New York, NY: Simon and Schuster, 2002).

[14] U.S. Centers for Disease Control and Prevention (CDC), 2004. Available at <http://www.bt.cdc.gov/agent/agentlist-category.asp>.

<u>CDC PRIORITY CATEGORIES</u>[1]

- *Category A: "The U.S. public health system and primary healthcare providers must be prepared to address various biological agents, including pathogens that are rarely seen in the United States. High-priority agents include organisms that pose a risk to national security because they can be easily disseminated or transmitted from person to person; result in high mortality rates and have the potential for major public health impact; might cause public panic and social disruption; and require special action for public health preparedness." Examples of agents currently in Category A: Anthrax (Bacillus anthracis), Botulism (Clostridium botulinum toxin), Plague (Yersinia pestis), Smallpox (variola major), Tularemia (Francisella tularensis), Viral hemorrhagic fevers, (filoviruses [e.g., Ebola, Marburg] and arenaviruses [e.g., Lassa, Machupo]).*
- *Category B: "Second highest priority agents include those that are moderately easy to disseminate; result in moderate morbidity rates and low mortality rates; and require specific enhancements of CDC's diagnostic capacity and enhanced disease surveillance." Examples of agents currently in Category B: Brucellosis (Brucella species), Epsilon toxin of Clostridium perfringens, Food safety threats (e.g., Salmonella species, Escherichia coli O157:H7, Shigella), Glanders (Burkholderia mallei), Melioidosis (Burkholderia pseudomallei), Psittacosis (Chlamydia psittaci), Q fever (Coxiella burnetii), Ricin toxin from Ricinus communis (castor beans), Staphylococcal enterotoxin B, Typhus fever (Rickettsia prowazekii), Alphaviruses (e.g., Venezuelan equine encephalitis, eastern equine encephalitis, western equine encephalitis), Water safety threats (e.g., Vibrio cholerae, Cryptosporidium parvum).*
- *Category C: "Third highest priority agents include emerging pathogens that could be engineered for mass dissemination in the future because of availability; ease of production and dissemination; and potential for high morbidity and mortality rates and major health impact." Examples of agents currently in Category C: Emerging infectious diseases such as Nipah virus and hantavirus.*

---

[1]CDC, 2002. For more discussion of the process to develop this list see "Report Summary: Public Health Assessment of Potential Biological Terrorism Agents," *Emerging Infectious Diseases* 8, no. 2 (2002): 225-230.

The categorization process also recognized that changes in microbiology, weaponization strategies, vaccination, technology, and other variables will lead to iteration, but the current priority list includes anthrax in Category A, and a recent report provides details about the NIAID's progress on its biodefense research agenda related to Category A agents.[15] With all of this focus on bioterrorism and recognition that any biological attack could create impacts on multiple scales (from local to international), the

---

[15] National Institute of Allergy and Infectious Diseases, *NIAID Biodefense Research Agenda for CDC Category A Agents – Progress Report*, August 2003. Available at <http://www2.niaid nih.gov/NR/rdonlyres/424B5936-3146-419D-BCF6-E5AA80045B4C/0/category_A_Progress_Report.pdf>.

Federal Government must recognizes its role in facilitating national biopreparedness. Providing consistency across all branches and levels of government, between the military and non-military government agents, and between all of the key stakeholders (including health care providers, nongovernmental organizations, industry, and the media) is an essential component of biopreparedness efforts. The challenge clearly emerges as one of successfully organizing enormous numbers of individuals and groups to ensure communication and cost-effective uses of resources—simply stated, finding a framework that facilitates all efforts to make good choices.

In the context of the wide array of potential agents, focusing on a single agent considerably narrows the set of issues. At the most basic level, several agent characteristics that affect the public health serve as criteria for sorting agents (e.g., into the CDC priority categories):

- **Treatability:** Currently we benefit from two major options for treating the health effects of biological organisms: pharmaceuticals (antibiotics, antivirals, etc.) and vaccines. We benefit from a large and diverse supply of antibiotics that effectively control many bacterial infections, but antibiotic resistance threatens the efficacy of these powerful tools.[16] Consequently, decisionmakers must consider the existing options for treating any disease, which may include vaccination and/or prophylaxis. This raises important questions related to developing a stockpile of vaccines and/or pharmaceutical products and the need for development of plans to use the stockpile when needed. If a vaccine exists, then decisionmakers face policy questions related to distributing the vaccine for different groups in the population (military, non-military, first responders, etc.).

- **Virulence:** Part of the assessment of priorities recognizes the differences between agents in their ability to produce human diseases. The combination of a susceptible population exposed to a virulent agent yields a big impact. The anthrax strain used in the October 2001 attacks differed significantly from the veterinary vaccine strain used unsuccessfully by Aum Shinrikyo, a Japanese cult that reportedly made several attempts to release *B. anthracis* without success in

---

[16] Kimberly M. Thompson with Debra Fulgram Bruce, *Overkill: How Our Nation's Abuse of Antibiotics and Other Germ Killers Is Hurting Your Health and What You Can Do About It* (Emmaus, PA: Rodale, 2002).

producing illness.[17] The virulence of an agent plays a critical role in decisions related to the choice and use of medications and communications to the public and authorities.

- **Environmental load and persistence:** The natural environment destroys many biological agents, but some agents show significant long-term viability (e.g., anthrax forms extremely hardy spores). The environmental load and persistence (a measure of the ongoing threat) determine the best strategy for managing an impacted area. Some of this depends on the organism used, but choices about management (cleanup, evacuation, etc.) depend on the concentration that people might be exposed to compared to the levels that cause illness. The exposure depends on the type of weaponization, the method of dispersion, spread, and processing techniques that impact the ability of the organism to effectively infect people. These issues lead to critical decisions about whether, how, and how much to cleanup, risk trade-offs between the agent and the often toxic chemicals used to clean, and risks to cleanup personnel.

- **Amplification potential:** Decisionmakers must consider several factors that might amplify the impact of an attack. First, a contagious disease (one that easily spreads through person-to-person contact) raises issues of protecting those initially not infected and questions about quarantine. (The treatability of the disease plays an important role and significantly impacts the choices.) Second, the public perception of the threat can significantly impact management; public health authorities must manage the chaos associated with panic[18] and deal with allocating limited treatment resources for those with and without illness.

---

[17] Amy E. Smithson and Leslie-Anne Levy, *Ataxia: The Chemical and Biological Terrorism Threat and the US Response* (Washington, DC: The Henry L. Stimson Center, October 2000), Report No. 35: xi. Available at <http://www.stimson.org/cbw/pubs.cfm?ID=12>.
[18] Thomas A. Glass and Monica Schoch-Spana, "Bioterrorism and the People: How to Vaccinate a City against Panic." *Clinical Infectious Diseases* 34 (January 15, 2002): 217-223.

Acting Wisely

While complexity makes management a challenge at all levels, the simplest concept of a framework is assessing the sets of decisions that must be made and fine tuned as information becomes available, and ensuring the creation of a process to answer these questions. The set of questions can be represented using decision trees, which policymakers can use to identify the key options and interactions between them and to understand the risks, costs, and benefits of choices. The following two pages represent a first attempt to identify a generic decision tree that includes major categories of decisions. The next section provides a first-cut example in the context of anthrax to demonstrate how the generic decision tree translates into an agent-specific tree.

RISK ANALYSIS AND DECISION TREES

At first glance, the use of decision trees seems a straightforward, even simplistic, approach to a very complicated problem. However, considerable analytical depth supports this approach. A brief overview of the field of risk analysis may help demonstrate its value.

Risk analysis includes risk perception, risk assessment, risk communication, and risk management, each of which may include multiple components. The process focuses on asking critical questions about the risks and understanding the uncertainties involved. With large uncertainties, risk analysis in this context of bioterrorism focuses on questions like: How probable is an attack? How lethal is a likely attack? What will be societal response? What can be done prior to an attack to limit its effect? After an attack?

Using decision trees provides a means of structuring the information related to choices clearly and efficiently. The Reverend Mr. Thomas Bayes, an 18th century mathematician and theologian who published his famous Bayes Theorem in 1763, receives credit for some of the earliest conceptual development. Remarkably, Bayesian analysis recently emerged as a powerful analytical tool with profound influence in the way we analyze decisionmaking.

By using a decision tree diagram, we can quickly see all of the options and combinations of paths, and the relationships may emerge as important considerations. This saves time and money when making decisions. By using such schemes now—especially during "wargames," drills, practices, etc.—much work can be done before an attack by identifying key pathways and preparing.

Bayesian networks also are useful because they basically "tell a story." A frequently used illustration includes five random variables: burglary, earthquake, alarm, neighbor call, and radio announcement to demonstrate the independence and conditional dependence of these variables.[1] The power of such an approach derives from the ability to take a collection of random, independent variables and easily calculate their joint probability.

---

[1] Daryle Niedermayer, "An Introduction to Bayesian Networks and their Contemporary Applications," December 1, 1998. Available at <http://www.niedermayer.ca/papers/bayesian/bayes.html>.

Generic Decision Tree

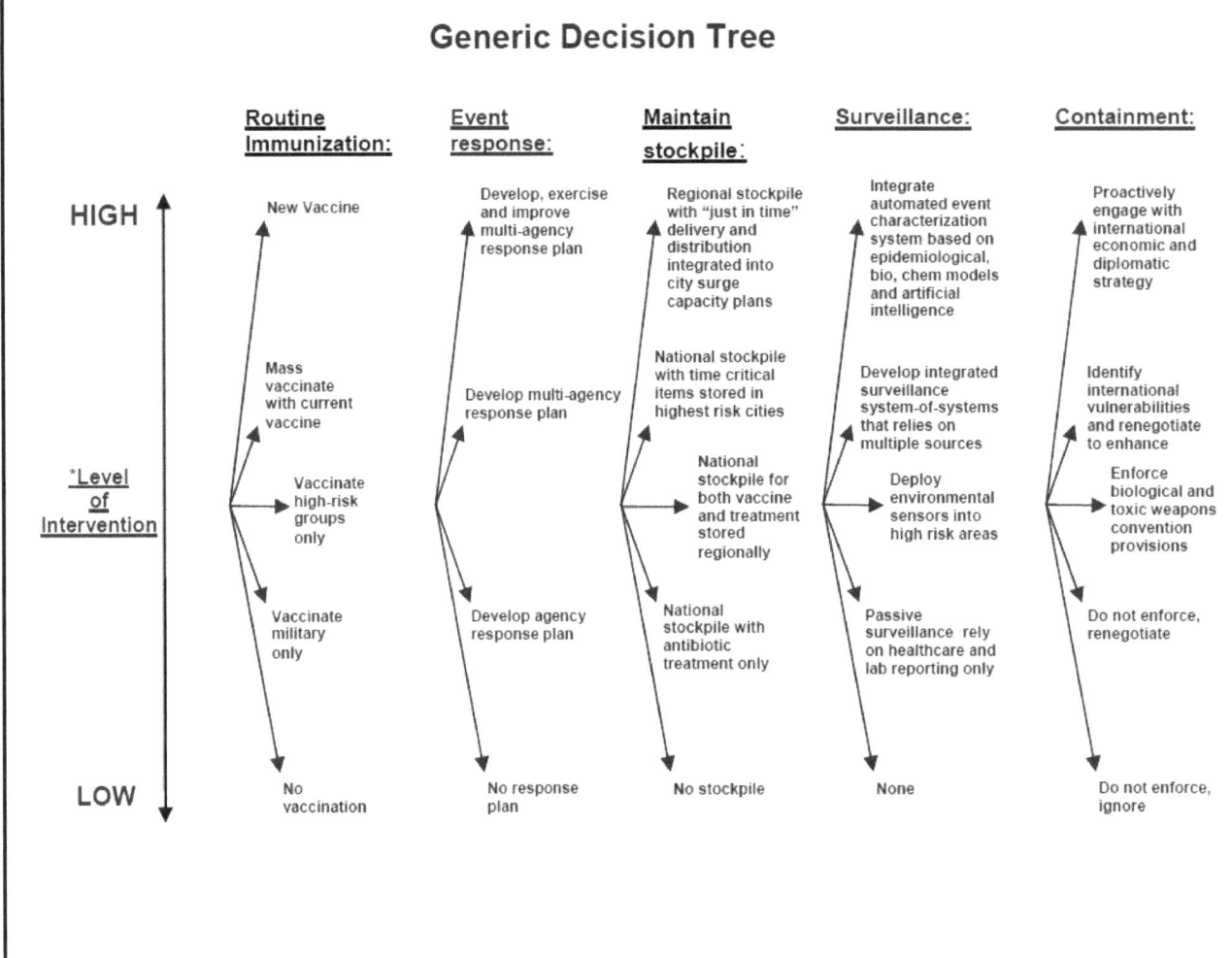

Figure 2 – The generic decision tree above can be used to develop a more explicit, purposefully tailored agent-specific tree for different pathogens or infectious diseases.

* "Level of intervention" refers to the financial costs incurred by the Federal Government for taking the initial path in any given decision tree. Hindsight, however, might reveal that low initial costs might mean higher long-term expenditures. (For example, no vaccination may seem an appropriate initial response in a hypothetical situation. Ultimately, however, the loss of life and subsequent compensation might prove it to have been a very costly choice.) The value of using such a standardized approach in wargames, drills, practices, etc., arises from the opportunity to consider many possible scenarios and examine their consequences at varying periods of time from the initial decision. Ultimately, this may lead to initial decisions that appear unnecessarily costly in the short run, but that may prove preferable as a result of the analysis of long-term consequences.

Generic Decision Tree

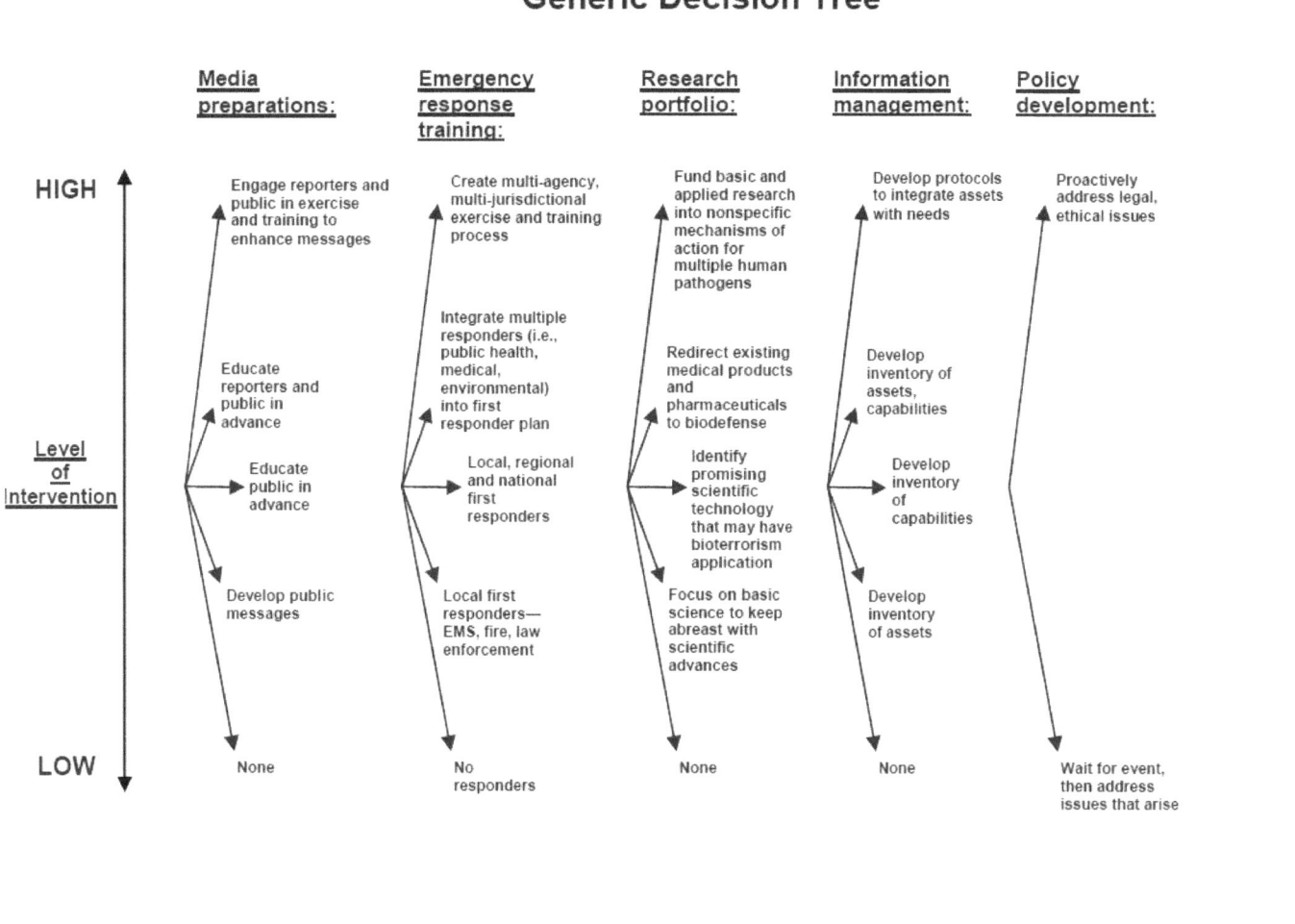

Figure 2 (cont.) – This represents a "first-cut" set of trees, as each initial decision then leads to a more refined tree for each topic.

The decision tree identifies 10 major categories of decisions that apply broadly for any potential biological weapon and that involve stakeholders at various levels and with divergent roles and responsibilities. Figure 2 also provides placeholders to demonstrate potential options intended to make the decision tree capture more of the complexity, but which should not be taken as the only or necessarily the appropriate options for any specific biological agent. We also recognize that other categories of decisions might emerge for some biological agents, and we expect iteration and improvement of the framework to result from the process of applying this approach to specific agents (like the example of anthrax in the next section). We elaborate on each of these generic decision areas to provide an overview of the decisions that we believe fall within each of these categories.

- **Routine immunization:** Fortunately, for many biological agents opportunities exist to prevent disease through immunization. Vaccines for endemic diseases generally represent highly cost-effective tools for managing those diseases and in some cases for significantly reducing the burden of the disease. However, in the context of potential bioweapons, decisionmakers generally face limited options related to vaccines. Consequently they must generally decide whether to seek development of a new vaccine for a specific biological agent. If a vaccine exists, then they must decide which groups of the population to vaccinate, if any, and evaluate the risks (mainly the adverse events of those vaccinated compared to the impact of having no one protected if an event occurs), costs (mainly comprehensive vaccine purchase and delivery costs), and benefits (mainly costs saved from better preparedness if an event occurs). This major decision category raises many potential sub-decisions about vaccination schedule, production constraints that generally involve the private sector vaccine producers, and vaccination policies (see the last category in the tree).

- **Event response:** This category represents a broad range of choices. At the most basic level it focuses on the questions related to developing and iterating on a comprehensive response plan for the specific biological agent. We recognize that no such plans currently cover all stakeholders and levels of government for any single biological agent, although some policymakers at some levels developed plans

12

covering their domains. Decisions under this category should include evaluation of the therapeutic options for treating people if an event occurs, which means assessing the cost-effectiveness of different antibiotics, antivirals, and other drugs that emergency responders might need if an event occurs. Clearly this leads to decisions related to rapidly acquiring the tools required to respond, including drugs and means to administer them (see the stockpile decisions that follow) and choices about restricting access and clean-up of impacted areas. It relates to decisions about surveillance, containment, media and risk communication preparations, and emergency response training (see these specific categories below). At this level, we assume that national policymakers must prioritize resources available for coordination efforts and make a choice about investing in a process to create an event response plan for a specific biological agent, creating and testing that plan, or not, although other options may also exist as indicated. Decisions here also raise a number of policy decisions (see last category of the tree).

- **Stockpile:** The CDC currently maintains a Strategic National Stockpile (SNS).[19] Thus, for any given agent, the key decisions should begin with assessments of the adequacy of existing resources, and include consideration of stockpiling vaccine (if one exists) and/or specific therapeutic agents as part of the SNS or separately (i.e., in an independent stockpile). Some decisionmakers may differ in the constraints on their choices. For example, military and non-military decisionmakers may evaluate the need for stockpiles differently given very different anticipated risks.

- **Surveillance:** Currently surveillance represents a "hot topic" given the proliferation of technologies developed to detect biological agents in people and the environment.[20] Decisionmakers often face a significant number of specific technological options for surveillance. At the baseline, no surveillance simply implies the current passive surveillance provided by the nation's health care

---

[19] Stephen D. Prior, *Who you gonna call? Responding to a medical emergency with the Strategic National Stockpile*, (Washington, DC: Center for Technology and National Security Policy, National Defense University, June 2004): 5.

[20] Robert E. Armstrong, Patricia K. Coomber and Stephen D. Prior, *Looking for Trouble: A Policymaker's Guide to Biosensing*, (Washington, DC: Center for Technology and National Security Policy, National Defense University, June 2004): 44-51.

providers, who detect and report disease following the presentation of patients with symptoms (as observed in the anthrax 2001 events). Other options exist to develop increased surveillance within the health care system, but still relying primarily on health care providers to collect information. Environmental surveillance represents another option, which could potentially detect organisms in the environment prior to patients presenting with symptoms. The critical questions that decisionmakers face in the context of investing in surveillance come down to understanding the value of the information obtained from surveillance: Do the benefits of the information obtained from any surveillance option under consideration exceed the system's costs (including the costs of false positives and false negatives)?

- **Containment:** The existing international agreements about biological agents, specifically the Biological and Toxic Weapons Convention Provisions (BTWCP), represented voluntary agreements to restrict development. If followed, they would preclude the need to address issues of containment. However, as demonstrated by the 2001 anthrax event and the still unknown perpetrator(s), those willing to use weaponized anthrax may have retained it, and this suggests the need for decisions about containment. For some biological agents, this decision may simply focus on whether and how to enforce existing policies, like the BTWCP, but for other agents it may involve negotiating new policy. To the extent that the government invests in research efforts related to the biological agent and this makes the agent more available (given its presence in research laboratories), this category also includes questions about policies for containment of the agent (and possibly related agents) in research labs.

- **Media and risk communication preparations:** Several studies demonstrate the lack of public knowledge about the risks of bioterrorism broadly and with respect to specific threat agents.[21] Given basic misunderstandings and the potential for miscommunication that could lead to costly and risky outcomes in the case of an event (e.g., traffic jams that prevent orderly evacuation, injuries and deaths from

---

[21] Baruch Fischhoff, Roxana M. Gonzalez, Deborah A. Small and Jennifer S. Lerner, "Evaluating the Success of Terror Risk Communications," *Biosecurity and Bioterrorism: Biodefense Strategy, Practice, and Science* 1 (2003): 255-258. Available at <http://www-marketing.wharton.upenn.edu/ideas/pdf/Small/Evaluating%20Biodef.pdf >. See also Blendon et al.

accidents resulting from panic), decisionmakers face important choices related to educating members of the media in advance so that they can more effectively communicate to the public and/or developing messages that they can roll out effectively in the case of an event. Given the general lack of risk education in the population, this area may also require an investment in research to understand mental models (perceptions, attitudes, and beliefs) and characterize anticipated behaviors.

- **Emergency response training:** All stakeholders recognize that emergency response may involve numerous groups ranging from local (since any event occurs at some locality) to international (e.g., exotic disease experts from around the world and global policymakers) and from health care to law enforcement. For any specific biological agent, decisionmakers face choices about which emergency responders to train, how to best train them, and how to facilitate the necessary coordination to ensure an optimal response to an event. Decisions may also include making choices about obtaining necessary personal protective equipment and communication tools for responders and training them to use these. Numerous table top exercises demonstrate the value of simulations for training purposes, but clearly lessons learned need to extend beyond the group involved in the actual exercise when possible. In the absence of a specific event response plan, emergency response training may still occur and still represents a set of decisions. If an emergency response plan exists, then it may call for or assume emergency response training, and decisionmakers may primarily face decisions related to allocating the resources necessary to achieve training objectives and evaluation of the training.

- **Research portfolio:** The choice to engage in research related to biological agents represents a spectrum of options including both basic science and social science. For any specific agent, identifying the key uncertainties that impact decisions is an important starting point. In the context of a vaccine-preventable disease, research efforts will be required to develop and test a vaccine. In some cases, basic science research that provides a better understanding of the organism itself might be important investments for the development of vaccines and therapeutic agents. Applied research, including risk analysis modeling (e.g., to

15

characterize the potential impacts of events), and social science research (e.g., to characterize people's behaviors and understanding of risks) represent critical research areas as well. This category represents the need for policymakers looking at the big picture to ensure a good match between research priorities that come from understanding the uncertainties that drive decisions and actual investments in research.

- **Information management:** For any specific agent, policymakers face choices about whether and how to manage the information. This includes basic types of information like inventories of places and people that possess certain capabilities and assets (including agents themselves). The existence of such information management systems also raises questions of security.

- **Policy development:** As discussed with many of the prior categories, policymakers must recognize that several legal and ethical issues arise in the context of managing the risks from biological agents. Issues like whether and how to quarantine, choices about providing and/or rationing vaccines and therapeutic agents in the case of an event, and sharing health information with potential violation of privacy protections in the interest of national security all represent obvious areas for developing policy positions and engaging in debate prior to an event. The decisions about developing policy should also serve to help decisionmakers understand existing policies and recognize constraints.

While we recognize that this framework initially provides just a starting point within each category, we also see critical insights that come from looking at managing the risks of a specific disease at this level. For example, while many of these observations may appear obvious, in some cases decisionmakers responsible for one part of the tree make assumptions that are inconsistent with the assumptions made by those responsible for other parts. Clearly, we cannot vaccinate if we haven't invested in the necessary research to develop the vaccine, which implies the need for alignment between research and immunization options. In the case of an outbreak, responders cannot treat effectively if they don't have enough of required therapeutics (if what they need is not widely available or is in an accessible stockpile). We cannot possibly detect an attack before it leads to adverse health effects if we aren't looking. Those responsible for management will not be effective in reassuring the public if they don't understand what the public knows and believes, or if the public doesn't trust them. It's also very important to recognize that we cannot assume that people will take the right action in response to an event if they don't know what that is.

Recommendations

- *Decisionmakers cannot effectively operate just in a limited scope when the choices of others significantly impact their options. We must engage in a process to develop a comprehensive national plan.*
- *HHS must develop a blueprint for a national biodefense strategy and a multiyear implementation strategy that includes relevant Federal, state, and local governmental organizations, private and public healthcare delivery systems, and industry.*

Understanding Anthrax

The October 2001 anthrax attacks, which killed 5 American citizens and cost billions of dollars to contain, decontaminate, and investigate, remarkably have not yet led to the development of a coherent national plan to manage the risks from the use of anthrax as a bioweapon. The absence of a coherent plan probably stems at least in part from the complexity of the various dimensions on which the risks must be considered, including the large number of stakeholders (e.g., local, state, national, and international levels of government and agencies, medical and public health communities, the media, law enforcement, military), the complexity of the science and assessment of the risks, and the potentially large resource requirements associated with managing these risks. As the mystery about the origin of the anthrax attacks continues to raise public skepticism about the ability of the government to deal with anthrax as a bioweapon, this is a critical time for the government to develop a national plan for managing the risks of anthrax.

Context

Historically, anthrax represents an important pathogen. Pioneers of microbiology worked on anthrax, including Pasteur and Koch, because during their lifetimes anthrax presented a significant, deadly disease in animals.[22] Based largely on their work and sustained efforts by veterinarians, anthrax was essentially eradicated from domestic animals in the last century. Consequently, anthrax disease in the U.S. occurs naturally only very rarely in humans, with most cases historically associated with animal contact. However, the anthrax letter attacks in 2001 made the threat of weaponized anthrax very real and suggest that containment of any existing anthrax remains an ongoing issue, and the fact that the criminals remain at large suggest an ongoing and real domestic threat. The attacks also demonstrated that weaponized forms of anthrax exist with demonstrated

---

[22] George Sternbach, "The history of anthrax," *Journal of Emergency Medicine* 24, no. 4 (May 2003): 464.

lethality,[23] even though they may present technical challenges that limit the ability of some potential attackers to produce them.[24]

More significantly, numerous opportunities exist to disperse anthrax, for example as an aerosol, in water[25] or food, or in solid media that travel (i.e., mail, packages, etc.). In addition, while antibiotics exist to treat people with anthrax when detected early, the technologies exist for attackers to create genetically modified, antibiotic resistant forms of anthrax (although this remains undemonstrated to date and the probability now seems low). Anthrax disease depends on contact with the agent in the environment, and not on person-to-person contact, a basic fact that one study found nearly half (47%) of the population surveyed did not know.[26]

---

[23] H. Clifford Lane and Anthony S. Fauci, "Bioterrorism on the Home Front," *Journal of the American Medical Association* 286 (November 28, 2001): 2595-2596. Available at <http://www.niaid nih.gov/director/pdf/jama_and_lane.pdf>.

[24] Thomas V. Ingelsby. "Anthrax: A possible case history," *Emerging Infectious Diseases* 5, no. 4 (July-August 1999): 556-560. Available at <http://www.cdc.gov/ncidod/EID/vol5no4/pdf/inglesby.pdf>; Thomas V. Inglesby, Donald A. Henderson, John G. Bartlett et al., "Anthrax as a Biological Weapon," *Journal of the American Medical Association* 281, no. 18 (May 12, 1999): 1736. Available at <http://jama.ama-assn.org/cgi/reprint/281/18/1735.pdf>.

[25] W. Dickinson Burrows and Sara E. Renner, "Biological Warfare Agents as Threats to Potable Water," *Environmental Health Perspectives* 107, no. 12 (December 1999): 975-984; W. Seth Carus, *Working Paper, Bioterrorism and Biocrimes: The Illicit Use of Biological Agents Since 1900* (Washington, DC: Center for Counterproliferation Research, National Defense University, 1998); Percy Frankland and Harry Marshall Ward, "Second Report to the Royal Society Water Research Committee: The Vitality and Virulence of *Bacillus Anthracis* and its Spores in Potable Waters," *Proceedings of the Royal Society of London* 53 (1893): 164-317.

[26] Fischhoff et al., 2.

WHAT IS ANTHRAX?

Anthrax is a bacterium with a long history of affiliation with the human race. Primarily, a disease of animals that eat vegetation—cattle, sheep, goats, camels, antelopes, etc., human exposure primarily comes from contact with infected animals or their tissue.

Humans can get infected through three routes:

1) Cutaneous (skin) exposure happens when the bacterium enters a wound on the skin, such as a worker handling contaminated animal wool or hides. About twenty percent of cutaneous infections will result in death, if untreated. Effective antibiotic treatments are readily available, and death of treated patients is rare.
2) Intestinal exposure results from eating contaminated meat. When untreated, death rates can be as high as sixty percent. However, effective antibiotic treatments make death unlikely.
3) Inhalational anthrax is usually fatal if untreated. Although it is curable with antibiotics, the time between exposure and onset of treatment needs to be very short—possibly as few as 36 hours.

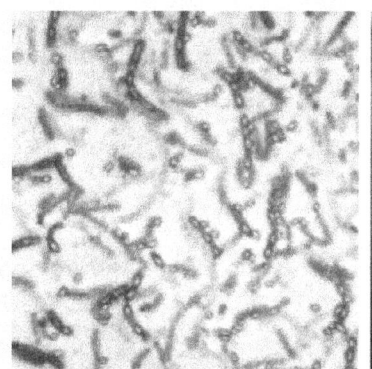 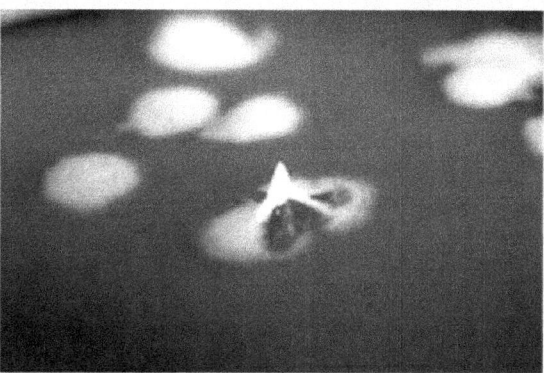

Figure 3 – Microscope picture of spores and vegetative cells of anthrax of the bacterium *Bacillus anthracis* (left) and colonies of the baterium *Bacillus anthracis* on blood agar (right). Source: WHO/Eric Miller, 2004, http://www.who.int/multimedia/anthrax/photo.html.

In general, anthrax is a problem mainly in developing countries that lack adequate veterinary public health programs. The fact that anthrax remains a powerful pathogen that exists in some places around the world makes it a sustained concern in the United States.

Several characteristics of anthrax may explain its use as the weapon of choice for the 2001 attacks. Anthrax survives as a very hardy organism that can form spores under severe environmental conditions. The exact environmental conditions that can trigger the bacterium to form spores remain uncertain, along with the conditions that cause the spores to revert to the bacterial form.

The warm, moist environment of the human lung, however, clearly offers an ideal environment for bacterial growth. Moreover, by entering the lung as spores, anthrax may effectively bypass the immune system's triggers and escape the first few lines of the body's defenses. By the time the immune system recognizes the presence of a foreign "invader," the attack is well underway.

Another aspect of the bacterium that makes it particularly well-suited to use as a bioweapon is its ability to produce a lethal toxin. (Most bacteria lack this capacity.) The toxin actually consists of three separate proteins, each harmless on its own. Working in conjunction with each other, however, the proteins gain entry into the body's cells and lead to death. Scientists continue to study the three collaborating proteins: protective antigen (PA), edema factor (EF) and lethal factor (LF) and to use knowledge about them to develop pharmaceutical products. Although antibiotics help to control the growth of the bacteria, they but do nothing to stem the effects of the toxin. Current research focuses on the development of drugs that would neutralize one or more of PA, EF, or LF. Such drugs could be used in concert with antibiotics to treat infected individuals.[1]

---

[1] John A. T. Young and R. John Collier, "Attacking Anthrax," *Scientific American* (March 2002): 48-59.

21

TOO MUCH OF A GOOD THING?

Antibiotics kill bacteria. That's a good thing. In the case of antibiotics, however, one can get too much of a good thing. In fact, the overuse of antibiotics and rise in antibiotic resistant bacteria have begun to threaten our ability to control common infections.

Antibiotics kill bacteria by disrupting activities within the bacteria.[1] For example, Cipro effectively destroys anthrax—and various other bacteria causing a variety of diseases from bronchitis to gonorrhea—by disrupting an enzyme that helps in the production of DNA. The anthrax cannot survive without the ability to reproduce, and it needs DNA to do so. The affected enzyme—topoisomerase II—does not exist in humans, so Cipro produces no effect on human cells.[2]

Over time, however, antibiotics become less effective because the bacteria adapt and change in one of three ways. Bacteria may change from a spontaneous mutation, or they may change through new genetic combinations that arise as a result of "transformation"—the equivalent of sexual reproduction between bacteria. Finally, small circles of DNA called plasmids can pass between bacteria and carry genes for drug resistance.[3]

As we introduce more antibiotics into the environment, this places greater selective pressure on the bacteria to evolve alternate enzymes or biochemical pathways that differ from those being acted on by the antibiotics. (Even if a person appropriately takes antibiotics, some still get introduced into the greater environment through elimination.) The development of antibiotic resistant forms of anthrax by the former Soviet Union provides a good illustration of how bacteria can change. Because of their rapid growth rate—meaning many generations produced in a short time-span—a Cipro-resistant strain eventually would arise from just normal, spontaneous mutation.[4] The development of such strains would not require any advanced genetic engineering techniques.

The problem of antibiotic resistant bacteria has reached the point where some consider it the number one public health problem.[5] Indeed, deaths from hospital acquired infections in the U.S. rose from 13,300 per year in 1992, to about 90,000 in 2004 because of the rise of antibiotic resistant bacteria.[6] Thus, needlessly taking Cipro—or any other antibiotic—as a prophylaxis against possible anthrax exposure contributes to an already serious problem.

---

[1] Bacteria are living organisms and are susceptible to the actions of antibiotics. Viruses are very different from bacteria and technically are not living organisms. Antibiotics are ineffective against viruses.

[2] Marshall Brain, "How Cipro Works." Available at <http://health.howstuffworks.com/cipro3.html>.

[3] Ricki Lewis, "The Rise of Antibiotic-Resistant Infections," *FDA Consumer Magazine* (September 1995): 12. Accessed at <http://www.fda.gov/fdac/features/795_ant bio.html>: *"In 1968, 12,500 people in Guatemala died in an epidemic of Shigella diarrhea. The microbe harbored a plasmid carrying resistance to four antibiotics!"*

[4] Ibid., 11.

[5] Helen Branswell, "Anthrax Scares May Fuel Growth of Ant biotic Resistance," *Canadian Press*, October 15, 2001. Available at <http://www.canoe.ca/Health0110/15_anthrax-cp.html>.

[6] National Institute of Allergy and Infectious Diseases, "The Problem of Antibiotic Resistance" (April 2004). Available at <http://www.niaid.nih.gov/factsheets/antimicro.htm>.

Anthrax disease has three forms: inhalational, gastrointestinal, and cutaneous, which differ in their characteristics. Inhalation of *B. anthracis* spores represents the major threat, because this leads to respiratory anthrax, the most lethal form of the disease, with mortality rates approaching 100% when untreated. Following the 2001 attacks, estimates of the dose that kills 50% of the population (i.e., the Lethal Dose 50, commonly written as LD50) decreased greatly as scientists learned more about the clinical character of inhalational anthrax. Today we remain uncertain about the minimum dose that can cause disease. Patients with inhalational anthrax present with nonspecific flu-like symptoms for a few days, and then suddenly show fever, dyspnea[27] (sometimes severe), diaphoresis,[28] hypoxia,[29] and shock, all of these rapidly progressing over a few hours. (Note that for any form, progression of the disease may depend on exposure and the nature of the agent used). Despite the lowered LD50, physicians learned that early recognition and aggressive treatment of symptoms greatly reduced death rates.

Gastrointestinal anthrax follows ingestion of *B. anthracis* spores from contaminated meat, which produces symptoms of nausea and vomiting, and leads to bloody diarrhea, sepsis,[30] and death.[31] The rarity of disease, combined with the inability to diagnosis it early, explains the observed high mortality rates in naturally occurring cases. Cutaneous anthrax produces characteristic lesions that clinicians can often detect early enough to effectively treat with antibiotics.

---

[27] Dyspnea is defined as difficulty in breathing, often associated with lung or heart disease and resulting in shortness of breath. This condition can also be referred to as "air hunger."
[28] Diaphoresis is another word for perspiration, especially when copious and medically induced.
[29] Hypoxia is classified as a deficiency in the amount of oxygen reaching body tissues.
[30] Sepsis is characterized by the presence of pathogenic organisms or their toxins in the blood or tissues.
[31] Thira Sirisanthana and Arthur E. Brown, "Anthrax of the Gastrointestinal Tract," *Emerging Infectious Diseases* 8, no. 7 (July 2002): 649-651. Available at <http://www.cdc.gov/ncidod/EID/vol8no7/pdf/02-0062.pdf>.

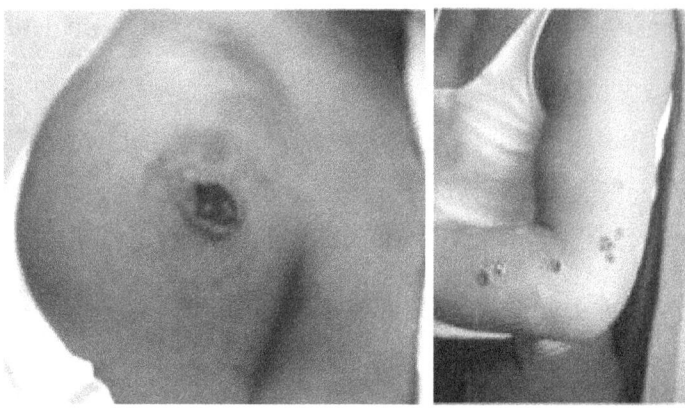

Figure 4 – Cutaneous anthrax lesions. *Source: CDC, 2004,*
http://www.bt.cdc.gov/agent/anthrax/anthrax/anthrax-images/cutaneous.asp.

Bioweapons experts recognized the potential threat of anthrax long ago and conducted research related to its weaponization. The long-term survival of spores, which the UK demonstrated can remain dangerous for decades in its tests at Gruinard Island,[32] makes anthrax a particularly costly agent, because it requires extensive cleanup. Investigations of the 1979 Sverdlovsk anthrax release in Russia demonstrate the lethality of aerosolized anthrax in humans,[33] and other studies provide insights into the potentially significant impacts of anthrax attacks in the U.S. and the potential benefits of early detection and response.[34]

Anthrax Attacks in October 2001

The anthrax attacks that occurred in 2001 provide important context for discussions about the U.S. management of anthrax risks. The attack involved envelopes containing weaponized anthrax sent on September 18 and on October 9, 2001 to

---

[32] British Broadcasting Company, "Britain's Anthrax Island," July 25, 2001. Available at <http://news.bbc.co.uk/1/hi/scotland/1457035.stm>.

[33] Jeanne Guillemin, *Anthrax: The Investigation of a Deadly Outbreak* (Berkeley, CA: University of California Press, December 1999); Matthew Meselson, Jeanne Guillemin, Martin Hugh-Jones et al., "The Sverdlovsk Anthrax Outbreak of 1979." Science 266 (December 9, 1994):1202-1208. Available at <http://www.anthrax.osd.mil/documents/library/Sverdlovsk.pdf>.

[34] Arnold F. Kaufmann, Martin I. Meltzer and George P. Schmid, "The Economic Impact of a Bioterrorist Attack: Are Prevention and Intervention Programs Justifiable?" *Emerging Infectious Diseases* 3, no. 2 (April-June 1997): 83-94. Available at < http://www.cdc.gov/ncidod/eid/vol3no2/kaufman.htm>. Also see Meselson et al., 1206-1207.

prominent government leaders and members of the media. The attacks ultimately led to 22 cases and 5 deaths.

Figure 5 – At the time of this publication's printing, the FBI's investigation into the Fall 2001 anthrax attacks, entitled "Amerithrax," was still ongoing. For more information, see *http://www.fbi.gov/anthrax/amerithraxlinks.htm.*

Figure 6 – Letter containing anthrax addressed to Senator Tom Daschle (D – SD). *Source: Federal Bureau of Investigation, http://www.thesahara.net/anthrax_rules.htm.*

Many uncertainties still exist related to these attacks, including the identity of the attacker and how several of the individuals exposed came in contact with anthrax. Several key lessons learned provide important insights related to planning for potential events:[35]

- Under-investment in the nation's public health infrastructure (which relies on mostly passive survey reports from doctors, uses the minimum possible workforce, and lacks coordination between public health agencies, emergency medical services, and law enforcement) meant that the system quickly became

---

[35] Heyman, 8-26.

strained by the attacks (with the least supported rural areas in some cases representing weak links[36]),[37]

- Insufficient laboratory capacity for processing samples, unsophisticated sample processing techniques, and lack of coordination within the laboratory network led to long delays in sample processing,[38]

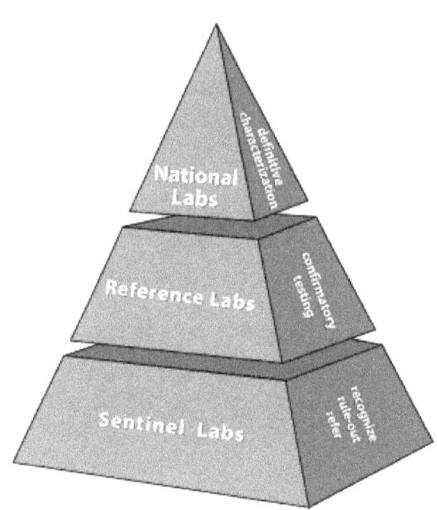

Figure 7 – A diagram of the Laboratory Response Network (LRN). The LRN includes 3 major types of labs: national, reference, or sentinel. The CDC and the U.S. Army Medical Research Institute for Infectious Diseases (USAMRIID) are the only two national laboratories, and they maintain responsibility for performing definitive characterization. Reference labs provide confirmation of samples from sentinel labs. Sentinel labs include the thousands of hospital labs in facilities that provide care to patients. These labs must recognize the suspicious samples and refer them to an appropriate reference lab. The labs all follow standard protocols for analyzing samples. The anthrax attacks of 2001 led to approximately 125,000 samples and more than 1 million separate tests by LRN labs. *Source: CDC, 2004, http://www.bt.cdc.gov/lrn/factsheet.asp* and *http://www.bt.cdc.gov/agent/LevelAProtocol/anthraxlabprotocol.pdf.*

- Inadequate knowledge of the pathology of disease caused by anthrax, characteristics of the bacteria, uncertain potency and effectiveness of treatment, and the lack of a national research agenda made it difficult to identify experts and rapidly obtain critical information,[39]

---

[36] Elin A. Gursky, *Hometown Hospitals: The Weakest Link?* (Washington, DC: Center for Technology and National Security Policy, National Defense University, June 2004): 1-4.
[37] Heyman, 8-9.
[38] Ibid., 10-11.
[39] Ibid., 11-12.

- Detection and surveillance relied on medical practitioners, some of whom lacked adequate training to respond appropriately (e.g., astute physicians who identified the probable threat and acted quickly to contain and warn other clinicians prevented a second death in Florida, but a Washington postal worker who later died of inhalational anthrax was sent home from a Maryland hospital with flu-like symptoms after a co-worker was admitted to another hospital with inhalational anthrax),[40]

- While responders took steps that helped to save lives, they did so without any plan, checklist, or clear list of best practices, and they lacked good information to make critical decisions about cleanup (e.g., how clean is clean enough and what methods clean the most cost-effectively?),[41]

- The lack of a clear chain of command slowed the process of management and cleanup, and the lack of prior training meant people involved in management did not already know each other or have a clear sense of responsibilities,[42]

- CDC's deployment of the antibiotics in the Strategic National Stockpile demonstrated the utility of this resource, but some challenges arose in the context of mass-medication and delivery. The experience also revealed the lack of strategies for dealing with the need for counseling large numbers of people, triaging and prioritizing patients for care and treatment, and potential issues of quarantine,[43] and

- Poor communication presented one of the largest problems. Media outlets sought statements from any potential expert, including uninformed spokespersons. Meanwhile, leading officials shared information poorly and generated conflicting messages.[44] The lack of a coordinated media strategy, combined with a scared public that was poorly educated about anthrax, led to many Americans seeking and using prophylaxis inappropriately, which further strained resources.

---

[40] Heyman, 14-15.
[41] Ibid., 24-25.
[42] Ibid., 16-19.
[43] Ibid., 19-20.
[44] Ibid., 20-23.

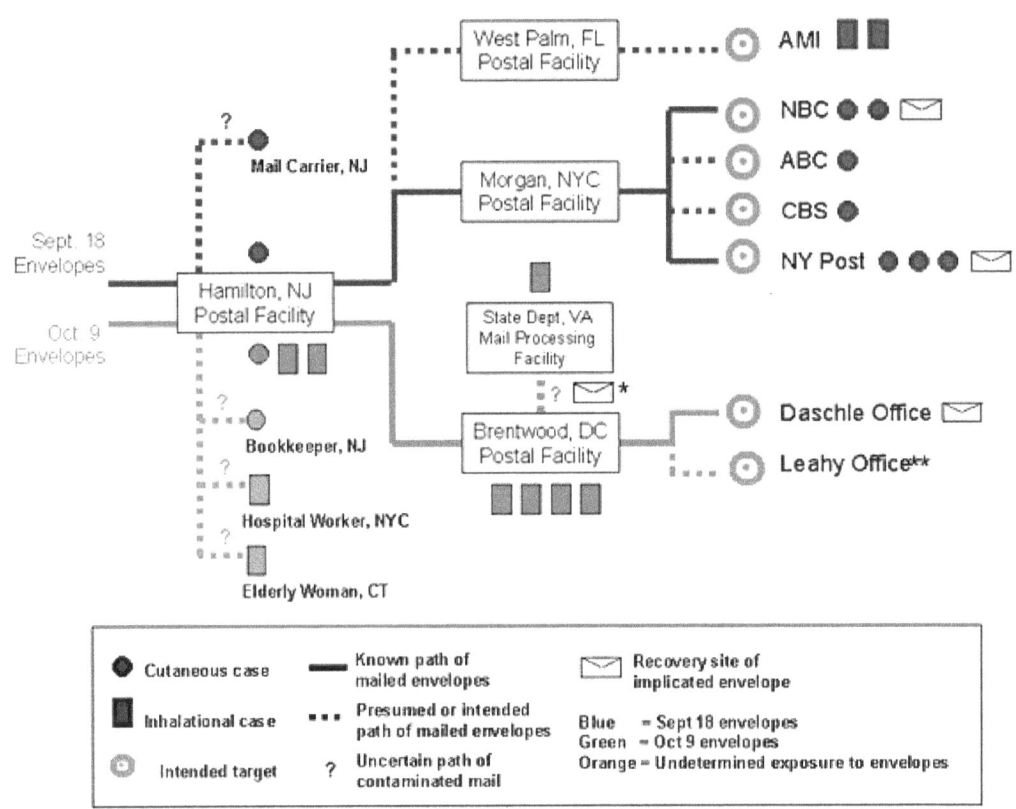

Figure 8 – Cases of anthrax associated with mailed paths of implicated envelopes and intended target sites. NY, New York; NBC, National Broadcasting Company; AMI, American Media Inc.; USPS, United States Postal Service; CBS, Columbia Broadcasting System. *Envelope addressed to Senator Leahy, found unopened on November 16, 2001, in a barrel of unopened mail sent to Capitol Hill; **dotted line indicates intended path of envelope addressed to Senator Leahy. *Source: Jernigan et al., (note that this is Figure 2 from the original article), supra note 1.*

Figure 9 – Images of FBI and EPA personnel sorting through sequestered Congressional mail and opening a letter addressed to Senator Leahy containing anthrax. *Source: http://www.fbi.gov/anthrax/vanharp/introleahy.htm and http://www.fbi.gov/anthrax/searchantpicts.htm.*

Managing Anthrax for the Future

This section develops a specific plan for managing any future anthrax incident and serves as a model for demonstrating the use of decision trees to help policymakers reach timely and science-based conclusions when faced with a public health crisis—whether caused by terrorists or Mother Nature.

The elements of a national plan for anthrax should address the wide range of issues identified in the generic decision trees. With respect to anthrax, while the risks remain uncertain, the 2001 attacks demonstrated the risks are real. The existence of a still unidentified attacker, potentially with access to weaponized anthrax, suggests that decisionmakers should actively consider the options that exist for managing them, the perspectives and roles of all stakeholders,[45] and the implications of their decisions in a broad context.

---

[45] The Business Roundtable, *Terrorism: Real threats. Real costs. Joint solutions* (Washington, DC: The Business Roundtable, June 2003): 13-18. Available at <http://www.businessroundtable.org/pdf/984.pdf>.

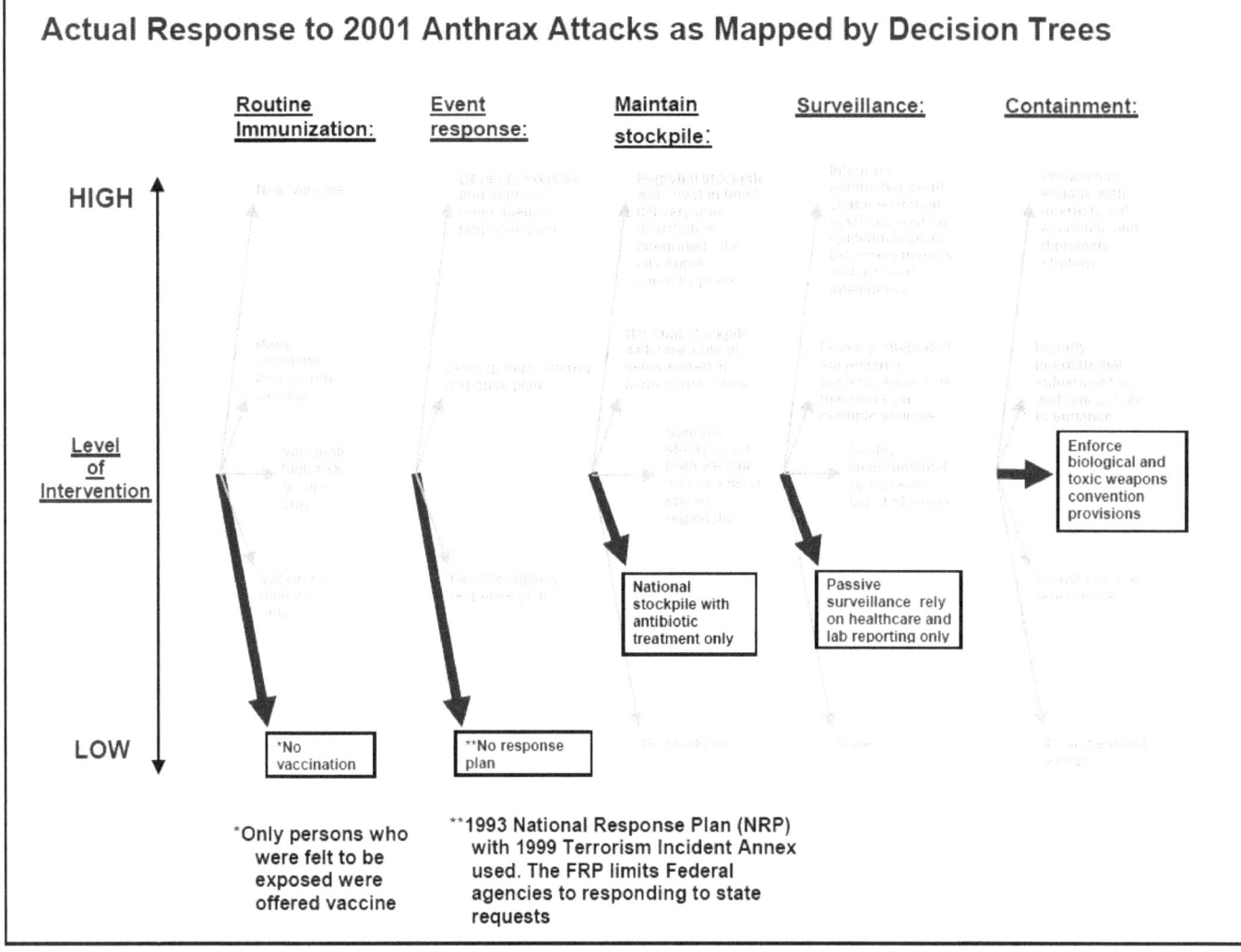

Actual Response to 2001 Anthrax Attacks as Mapped by Decision Trees

Figure 10 – This decision tree map of the 2001 anthrax attacks was developed after the fact, using the analysis of the incident conducted by Gursky, et al. (Gursky, E, Inglesby, T., and O'Toole, T. 2003. "Anthrax 2001: Observations on the Medical and Public Health Response." Biosecurity and Bioterrorism: Biodefense Strategy, Practice, and Science, 1:97-110).

These decision trees map the actual response taken by the Federal Government in 2001. Note that, in general, the relatively conservative initial responses ranked as "low" with respect to level of intervention. However, some of the low-level responses ultimately resulted in costly consequences. See pages 32-33 for a proposed set of responses in the event of future anthrax attacks.

Actual Response to 2001 Anthrax Attacks as Mapped by Decision Trees

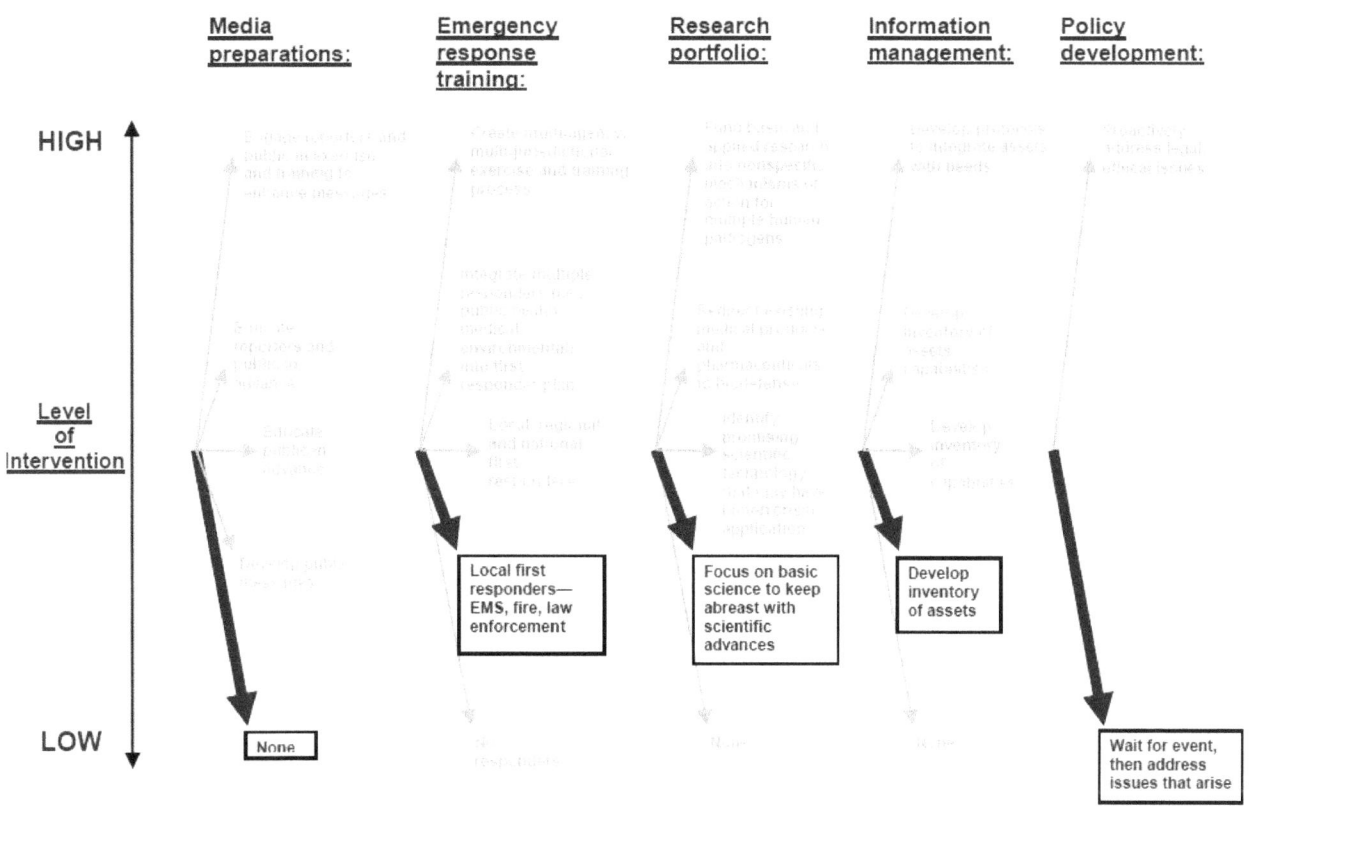

Figure 10 (cont.)

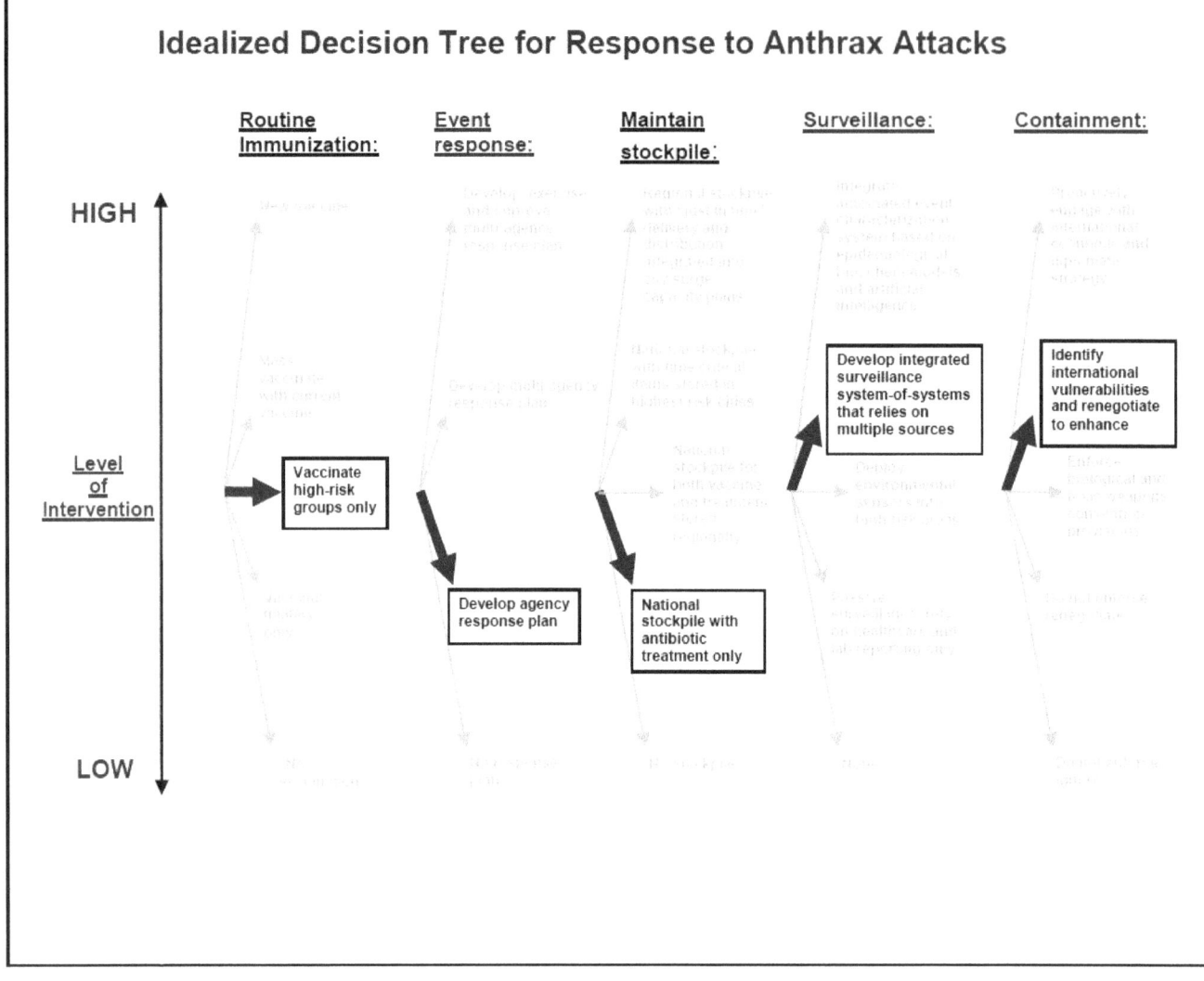

Figure 11 – The first-cut decision tree, tailored to more effectively manage anthrax.

This figure shows the recommended first-cut trees for anthrax. The remainder of this section discusses each of the ten trees in the context of existing policies that provide important context for future management. While quantifying the risks remains a challenge, and the science and technology continue to evolve, we believe that organizing the information in this way provides a very useful tool for integration and communication. We also review existing policies to provide context for current delegations of authority/responsibility to exercise the different options.

Idealized Decision Tree for Response to Anthrax Attacks

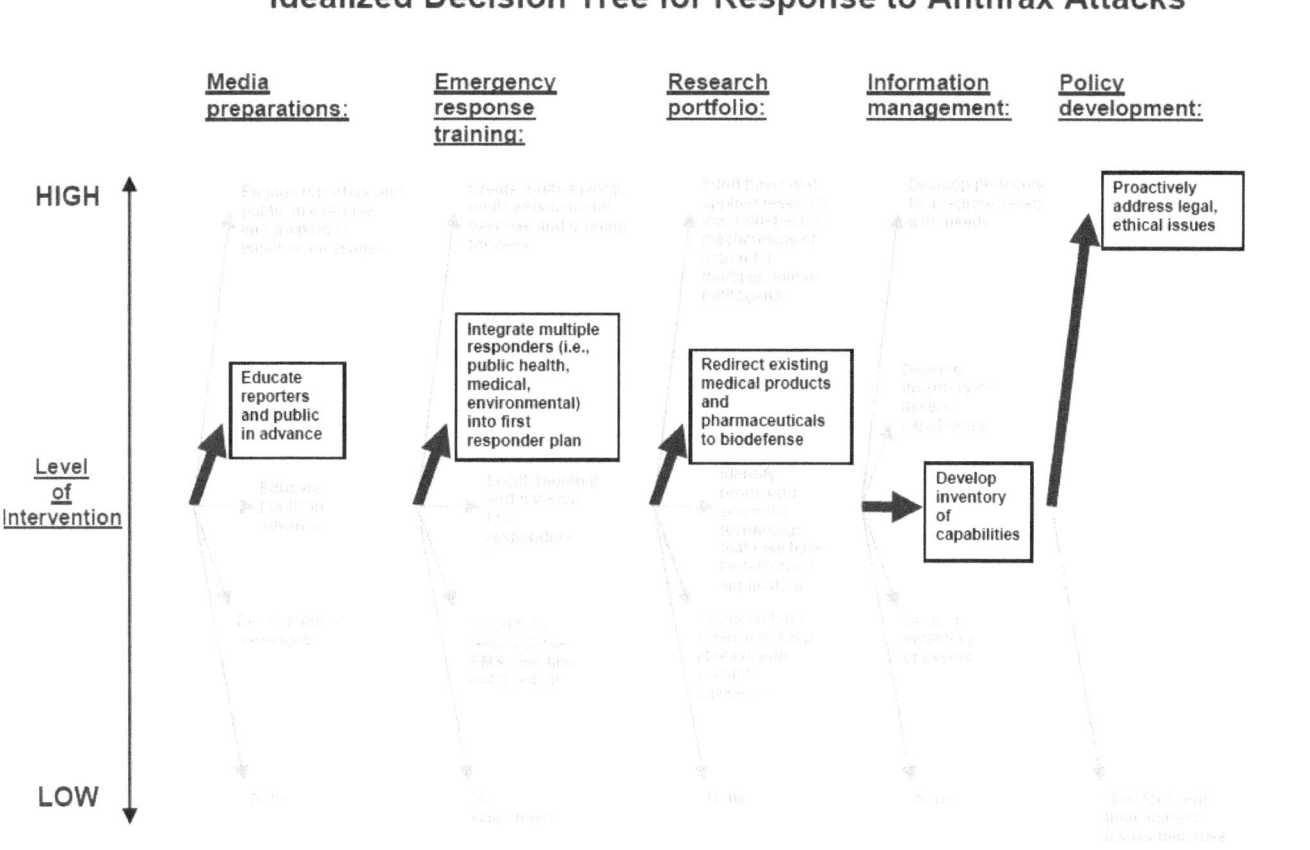

Figure 11 (cont.) – The first-cut decision tree, tailored to more effectively manage anthrax.

**These recommendations are meant not to criticize the actual handling of the 2001 anthrax attacks,** but to illustrate the benefit of careful consideration of the long-term outcomes of decisions. These recommendations clearly should remain the subject of debate and benefit from three years worth of analysis and review of the actual events. The value of using decision trees during wargames, practices, drills, etc., comes from their ability to provide a de facto analysis of the long-term consequences of decisions made during the early stages of an event.

Routine immunization

The U.S. first licensed a human vaccine for anthrax in 1970 using a non-encapsulated, toxigenic strain that reduces the virulence of the bacterium. The FDA approved Anthrax Vaccine Adsorbed (AVA) for veterinarians, workers handling potentially infected animals or their products, and laboratory workers conducting research on anthrax. When exposed to vaccine, the body produces an antibody response to protective antigen, one of three proteins produced by anthrax as part of its toxin. (See the "What Is Anthrax?" text box on page 21.) The existing dosage schedule for effective protection pre-exposure involves 6 doses given subcutaneously over 18 months followed by an annual booster. The vaccine poses risks of systemic reactions and serious side effects similar to flu and hepatitis, and local reactions in the arms of 30-60% of recipients, with women showing higher rates of local reactions than men. It is not indicated for pregnant women and must be given with caution to immunosuppressed persons. Studies of the vaccine conducted in 1962 demonstrated its effectiveness for workers in wool mills, primarily against cutaneous exposure to anthrax.[46] Although the military maintained high interest in using AVA to protect troops, limited and unreliable supply of AVA and issues related to the limited license that did not include aerosol exposure presented challenges. More recent tests in rhesus monkeys show effectiveness of AVA against inhalational anthrax,[47] but the military faced significant challenges to its policy of mandatory vaccination for anthrax, and controversy continues to date, even though surveillance shows few, if any, clinically significant side effects.[48] (See the "Anthrax Vaccine" text box on page 37.)

---

[46] Philip S. Brachman, Herman Gold, Stanley A. Plotkin et al., "Field evaluation of a human anthrax vaccine," *American J Public Health* 52, no. 4 (April 1962): 632-645. Available at <http://www.anthrax.osd.mil/media/pdf/field_eval.pdf>.

[47] Arthur M. Friedlander, Phillip R. Pittman and Gerald W. Parker, "Anthrax Vaccine: Evidence for Safety and Efficacy Against Inhalational Anthrax." *JAMA* 282 (1999): 2104-2106. Available at <http://jama.ama-assn.org/cgi/reprint/282/22/2104.pdf>.

[48] Jeffrey L. Lange, Sandra E. Lesikar, Mark V. Rubertone and John F. Brundage, "Comprehensive Systematic Surveillance for Adverse Effects of Anthrax Vaccine Adsorbed, US Armed Forces, 1998-2000," *Vaccine* 21, no. 15 (April 2, 2003): 1620. Available at <http://www.anthrax.osd.mil/documents/library/science.pdf>.

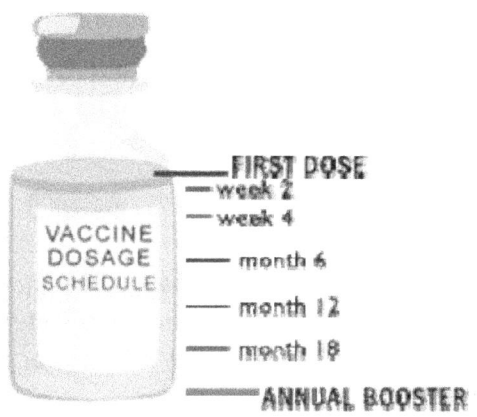

Figure 12 – Anthrax vaccine adsorbed (AVA) dosage schedule. *Source: http://www.anthrax.osd.mil/vaccine.*

Currently, no routine, non-military, non-occupational, human immunization for anthrax occurs.[49] A practical universal vaccination option does not exist, given the severely limited supply of vaccine and unclear public receptivity to mass vaccination for anthrax.

Research efforts continue to improve the scientific knowledge about anthrax and show promise toward the development of a new vaccine.[50] The NIAID commissioned VaxGen to begin human clinical trials on a new, and hopefully safer anthrax vaccine, rPA102 (see the "Anthrax Vaccine" text box on page 37) that the U.S. Army Medical Research Institute of Infectious Diseases (USAMRIID) previously developed and tested in animals.[51] This vaccine remains several years from licensure and large-scale production potential, but research efforts to support expanded routine immunization options are underway, and these may lead to future discussions about potential mass immunization.

---

[49] Advisory Committee on Immunization Practices (ACIP), "Use of Anthrax Vaccine in the United States," *MMWR Recommendations and Reports* 49, no. 15 (December 15, 2000): 1-4. Available at <http://www.cdc.gov/mmwr/preview/mmwrhtml/rr4915a1.htm>. Supplement available at <http://www.cdc.gov/mmwr/preview/mmwrhtml/mm5145a4.htm>.

[50] Gi-Eun Rhie, Michael H. Roehrl, Michael Mourez et al., "A dually active anthrax vaccine that confers protection against both bacilli and toxins," *PNAS* 100, no. 19 (September 16, 2003):10925-10930. Available at <http://www.pnas.org/cgi/reprint/100/19/10925>.

[51] VaxGen, Inc. website, available at <http://www.vaxgen.com/products/index.html>.

Based on current policy, we expect the U.S. to continue its use of routine vaccination only for high-risk groups, which means that most of the population remains vulnerable to a large-scale attack.

Recommendations

- *Given the possibility of a new vaccine in the future, efforts should be initiated now to determine the public acceptability of a new vaccine as a function of anticipated risks and benefits, and cost-effectiveness studies should evaluate how investment of public health resources in routine vaccination compare to other competing options.*
- *More broadly, with public health resources already scarce, the government should evaluate the ability to increase overall public health expenditures to support the inclusion of additional vaccines.*
- *HHS should consider alternate, risk-based vaccine and prophylaxis strategies where first responders (emergency medical services, fire fighters, and law enforcement) as well as emergency room doctors and staff are provided treatment. (In the case of smallpox, the treatment of first responders is especially important to consider.)*

ANTHRAX VACCINE

The anthrax vaccine generated significant controversy involving Federal judges, Congressional hearings, government regulatory agencies, National Research Council study committees, and "disobedient" service members.

The anthrax vaccine—known as Anthrax Vaccine Adsorbed (AVA)—received its license in 1970. The AVA manufacturers initially targeted veterinarians and workers who processed animal products, such as hair or hides that might harbor anthrax spores. Use of the vaccine expanded to the Department of Defense (DOD) in the 1990s. Some 150,000 of the military personnel deployed to the Gulf War received AVA, and in 1998 the Anthrax Vaccine Immunization Program (AVIP) called for the vaccination of all military personnel.

The AVIP effort slowed by 2000 as a result of limited vaccine supply and delays in Federal approval for the release of newly manufactured vaccine lots. Concerns expressed about AVA's efficacy and safety, as well as its production, contributed to the remarkable result that some 500 service members refused vaccination under AVIP and became subject to disciplinary action, including discharges, fines, and jail sentences.[1])

As a result of the concerns, Congress directed the conduct of an independent analysis, and in October 2000, the Institute of Medicine convened the Committee to Assess the Safety and Efficacy of the Anthrax Vaccine. The committee concluded that AVA was an effective vaccine. Moreover, because part of the vaccine's action was directed against the bacterium's toxin, the committee noted that the vaccine should be effective against all known strains of the bacterium, as well as any potential bioengineered strains.[2]

AVA was also deemed to be reasonably safe. The committee noted accounts of reactions, but concluded they were within the guidelines associated with any vaccine.

The production procedures at BioPort, the country's only manufacturer of AVA, were also reviewed by the committee. With newly validated manufacturing processes being approved at BioPort's renovated facility, the Food and Drug Administration (FDA) anticipated a greater assurance of vaccine consistency.

At the end of October 2004, however, a Federal judge ordered the DOD to stop requiring vaccination of all military personnel. In response to the order, the Secretary of Defense ordered a "pause" in the program. (The judge's ruling was based on his finding that the FDA did not follow its own rules requiring public comment on the vaccine's safety and effectiveness.) The military uses a layered approach to protection, so "pausing" the vaccination program will not leave troops completely unprotected.[3]

A new version of the vaccine is set to be manufactured by VaxGen, Inc., a California-based company. The vaccine is manufactured differently from the one made by BioPort and is based on the bacterium's own Protective Antigen (PA), one of the proteins involved in the production of the anthrax toxin. (The new vaccine is made with recombinant DNA technology and is called rPA102.) Vaccination with rPA102 will require only three shots, vs. the current six. Because it is based on a single protein, it potentially should have fewer side-effects.[4]

The new vaccine is being purchased by the Federal Government under the Project Bioshield Act of 2004. Under that legislation, the Department of Health and Human Services is seeking to secure a stockpile of 75 million doses of anthrax vaccine for use by the civilian population.

---

[1] "Judge Halts Forcing of Anthrax Shots," *New York Times*, October 28, 2004. Available at <http://www.nytimes.com/2004/10/28/politics/28anthrax.html>.

[2] Lois M. Joellenbeck, Lee L. Zwanziger, et al. (eds.), *The Anthrax Vaccine: Is It Safe? Does It Work?* (Washington, DC: Committee to Assess the Safety and Efficacy of the Anthrax Vaccine, Institute of Medicine, National Academies of Science, 2002): 71.

[3] Ibid., 33.

[4] VaxGen, Inc. corporate press release, October 15, 2004, available at <http://www.vaxgen.com/pressroom/index.html>.

DISTRIBUTING THE DRUGS: WHERE TO START?

Pharmaceutical intervention—whether vaccines, antibiotics, or antivirals—will likely present a difficult problem for policymakers, regardless of the agent used in a bioattack. This table presents a proposed scheme for prioritizing the use of available drugs, given the high probability that supplies will be limited or, at best, not immediately available in the affected area. (Note the middle box at the bottom of the table for specific references to anthrax.)

Category 1 - (Highest priority) (public health recommendation)
- Persons known to be exposed or at the geographic location of exposure
- Persons with clinical symptoms consistent with disease

Category 2 - (public health recommendation)
- Household or work contacts of known or probable cases
- Healthcare and first responder personnel with known or probable contact with cases

Category 3 - (public health recommendation)
- All healthcare, law enforcement, first responder, and public health personnel in geographic location of release or cases
- Potentially exposed persons based on contact tracing

Categories 4, 5 and 6 apply when resources are inadequate for mass prophylaxis/vaccination

----

Category 4
- Essential service personnel, e.g., Continuity of Government (COG)(government officials), public works personnel, private security

Category 5
- Utilities, transportation, communications workers

Category 6
- Home healthcare workers, consider community volunteers, high priority commercial workers (food, financial), news media

Potential initial prophylaxis strategies

Communicable agents – smallpox, plague

Attack parameters	Exposure Categories for Prophylaxis
Unknown source	1, 2, 3
Small, confined, localized area	1, 2
Large area, many exposed persons	1, 2, 3
Multiple sites or subsequent releases	1, 2, 3, consider 4 in severe circumstances
Secondary transmission beyond release point	Add 1-2 categories

Anthrax (non-communicable; contacts should not be treated. Potential exposure groups may need to be broadened, as symptoms may not start until 4-6 days after exposure)

Attack parameters	Exposure Categories for Prophylaxis
Unknown source	1, 3, consider 4
Small, confined, localized area	1, 3
Multiple sites or subsequent releases	1, 3, consider 4 and 5

Tularemia, Ricin, Botulinum toxin (non-communicable; contacts should not be treated)

Attack parameters	Exposure Categories for Prophylaxis
Unknown source	1, 3
Small, confined, localized area	1, 3
Multiple sites or subsequent releases	1, 3, consider 4

Event response

Based on the 2001 attacks, the CDC developed many components of a comprehensive event response plan. For example, the CDC provides a dedicated page on its Internet site for anthrax[52] that provides links to its antibiotic treatment guidelines for clinicians,[53] protocols for laboratory testing and confirmation,[54] epidemiological tool kit,[55] sampling and cleanup guidelines,[56] and other resources.

With respect to treatment, the CDC guidelines suggest treatment with antibiotics that are readily available because of their use to treat other infectious diseases and are part of the Strategic National Stockpile (e.g., Cipro and doxycycline). The CDC also tested post-exposure vaccination during the 2001 anthrax attacks under an Investigational New Drug (IND) application that used informed consent.[57] The post-exposure dosage schedule involved 3 doses given subcutaneously over 4 weeks along with 60 days of antibiotic. The scientific literature currently lacks studies of estimates of the cost-effectiveness of AVA, other potential vaccines, or antibiotics for treatment of anthrax, or papers that evaluate the cost-effectiveness of prevention and preparedness efforts.

Cleanup remains an important area of event response that warrants additional attention, particularly given the magnitude of the resources required. While the National Contingency Plan provided a means for the EPA to use part of the Superfund for cleanup in the Hart Senate Office Building, the U.S. Postal Service (USPS) cleanup and the American Media Inc. (AMI) building proved much more difficult. The USPS faced widespread contamination of a very large area, and contractors faced significant technological challenges in designing and implementing a cleanup strategy. In 2003, Congress bought the AMI building, which remained closed after the death of one of the

---

[52] CDC, 2004, available at <http://www.bt.cdc.gov/agent/anthrax/index.asp>.
[53] CDC, 2004. available at <http://www.bt.cdc.gov/agent/anthrax/anthrax-hcp-factsheet.asp>.
[54] CDC, 2004. available at <http://www.bt.cdc.gov/agent/anthrax/lab-testing/index.asp>.
[55] Dori B. Reissman, Ellen B. Steinberg, Julie M. Magri and Daniel B. Jernigan, "The Anthrax Epidemiologic Tool Kit: An Instrument for Public Health Preparedness," *Biosecurity and Bioterrorism: Biodefense Strategy, Practice, and Science* 1, no. 2 (2003):111-116. Available at <http://bioterrorism.dhmh.state md.us/AnthraxEpiToolkit.pdf>.
[56] CDC, 2004, available at <http://www.bt.cdc.gov/agent/anthrax/environment/index.asp>.
[57] CDC, 2001. available at <http://www.bt.cdc.gov/DocumentsApp/Anthrax/12182001/hhs12182001.asp>.

workers.[58] Issues of quarantine for contagious agents in event response, and restricting access and cleaning up private property remain significant ones that still need attention. Such efforts will require engaging the business community effectively.[59]

Recommendations

- *Efforts to evaluate the benefits of the substantial national governmental investments in bioterrorism preparedness deserve attention, particularly if the investments provide dual-use benefits or impose costs by diverting resources from other public health priorities. The nation needs a strategy to ensure that all levels of responders are ready in the event of another attack with anthrax, and this requires a sustained commitment and discussions with even broader groups of stakeholders.*
- *HHS should perform a comprehensive assessment of biodefense grants to identify and prioritize those that both meet urgent preparedness needs and provide collateral benefit to the public health infrastructure.*

Maintaining a stockpile

Currently, the SNS contains the antibiotics that authorities would need to respond to an attack with anthrax, and efforts to procure vaccine are underway. CDC deployment of the SNS represents one part of the system working in the 2001 anthrax attacks,[60] and the future decisions focus on maintaining this resource.

Recommendations

- *Analysts should construct mathematical models to estimate the potential size of attacks and compare these to the stockpile resources. Various designs for the stockpile could be developed. In addition, analysts should consider whether the inclusion of some pharmaceutical products in the SNS might create incentives for attackers related to bioengineering agents resistant to SNS available treatments, and whether (and how) the actual formulary of the SNS should periodically change.*
- *HHS/CDC should reevaluate SNS contents and distribution processes and develop enhanced mechanisms to deliver those few, time-sensitive items to the US population at the greatest risk. In addition, policies should be developed— based on mathematical modeling—for deploying SNS resources if supplies are limited.*

---

[58] Kathy Bushouse, "Congress agrees to buy, clean up anthrax-tainted AMI building," *South Florida Sun-Sentinel*, February 14, 2003. Available at <http://www.ph.ucla.edu/epi/bioter/cleanupamianthax.html>.
[59] The Business Roundtable, 2003, i.
[60] Heyman, 3.

Surveillance

Many options exist to deploy sensors that might assist in providing warning of an attack, but these systems are currently under development and they raise many issues.[61] While it appears that efforts to develop new technologies abound, the government has not focused sufficient attention on important questions about the value of the information that might be obtained from these systems. At the broad level, system designers must determine whether anthrax is something for which they should be looking, presumably because an early warning might prove valuable in saving lives or response costs, or because it allows capture of the attacker and possibly retribution. However, surveillance can yield both false negatives, meaning that it can miss real cases, and false positives or false alarms, meaning that it incorrectly indicates an attack.

As promising technologies emerge, they should be evaluated with respect to the quality and cost of the information that they provide. If the information does little to speed up response or could lead to costly false alarms, which may include some injury-related deaths if people panic or receive some inappropriate treatment, then careful consideration of the trade-offs should drive decisions about deployment. For any proposed technology, decisionmakers must play a critical role in discussions about how they would use information, particularly imperfect information. Efforts to characterize the full costs of these systems, including the costs of training the people who might receive information from them, warrant particular attention in informing the national debate about biodefense.

---

[61] Armstrong, 6-16.

Recommendations

- *In the case of anthrax, authorities should evaluate the value of the information from a sensor network and compare this to the cost of such a network. While no technology currently exists for wide-scale deployment, efforts to develop the analytical methods needed to model the sensor system and the uncertainty in the information it might provide are a priority. These methods should consider the general guidance on cost-effectiveness analysis[1] and focus on the incremental benefits of any new technologies over the existing system of passive surveillance by the health care system. They should consider the training of health care providers as part of this analysis.*
- *HHS should commission a study that integrates governmental, civilian, and military needs and capabilities to develop a prioritized national action plan for surveillance and response.*

[1] Marthe R. Gold et al., *Cost-effectiveness in Health and Medicine*, (New York, NY: Oxford University Press, 1996).

Containment

The existing Biological and Toxic Weapons Convention Provisions provide agreements about biological agents, but compliance with these agreements and their enforcement remain on-going challenges. The fact that the 2001 anthrax attacks appeared to use anthrax developed in the U.S. raises important questions about containment of any existing anthrax here and abroad. The CDC is undertaking an effort to develop an inventory for anthrax entitled the Select Agent Program,[62] a first step toward national containment. However, important decisions remain about our international agreements, and given the existence of bioweapons as justification for its decision to go to war in Iraq, the government faces complicated policies and politics.

In addition, the expanded research on biological agents motivated by the anthrax attacks and significant infusion of resources has expanded the market for these agents. In the context of vaccine development and other scientific studies, researchers sometimes require access to agents. These materials must be produced and transported, which creates opportunities for breaches of containment. Recently, the Children's Hospital and Research Center at Oakland reported a serious breach of *B. anthracis* containment that

---

[62] Learn more about the CDC's Select Agent Program at <http://www.cdc.gov/od/sap/>.

led to exposure of 7 lab workers to live *B. anthracis* instead of inactivated bacteria. (All of the researchers took a 60-day course of Cipro and the material was recontained.[63])

<u>Recommendations</u>

- ***The fact that these decisions are complicated suggests the need for particular attention to resources and policy tools that promote destruction and containment of potential bioweapons. The U.S. government successfully removed from the open patent literature some information that could prove useful, but the rapid evolution of biotechnology means that containment efforts must address molecular synthesis capabilities and the existence of highly-trained and well-equipped scientists who continue to advance our knowledge. Containment efforts should ensure that all involved with hazardous biological agents receive adequate training and face incentives that encourage compliance.***

Media preparations

In the 2001 anthrax attacks, communication was one of the weakest links, and little has been done to improve the situation. The reality of bioterrorism risks means that authorities should place higher priority on risk communication and risk education to avoid panic and achieve the optimal response.

Studies of public opinion and perception of biological weapons suggest that many Americans lack basic knowledge about pathogens and how they cause disease. The fact that a study found that 47 percent of the population surveyed did not know that anthrax does not spread person-to-person suggests that, despite all of the media attention on anthrax from the 2001 attacks, people still need education about basic anthrax information.[64] Additional studies may also reveal other misperceptions about anthrax.

---

[63] John Dudley Miller, "US lab is sent live anthrax," *The Scientist*, June 11, 2004. Available at <http://www.biomedcentral.com/news/20040611/03>.
[64] Fischhoff et al., 255.

Recommendations

- ***Clearly we need to engage in active risk communication efforts to better prepare the country to manage risks and to ensure that all stakeholders recognize their important roles. Efforts to train members of the media to communicate uncertainty and to provide context will help, but identifying spokespeople and developing messages before they're needed are important priority areas in which work should begin now.***
- ***HHS and CDC—in conjunction with private sector healthcare clinicians—should engage media and risk communication experts to proactively develop effective communication strategies that consider message, messenger, and recipient.***

Emergency response training

A bioterrorism event involves people at all levels of government. The need for a response plan is clear, yet national investments in training must focus on developing cost-effective strategies for insuring preparedness while respecting the limits of time and other resources. The CDC invested $1.1 billion in 2002[65] and $1.4 billion in 2003[66] to assist state and local governments in the development of bioterrorism response plans, but remarkably little coordination of these efforts has occurred between states. More significantly, the effort does not include assessment of common needs or development of generic models that might assist in characterizing the spread of disease or impact of response options. More than two years after the program began, national progress has been inadequate.

With respect to the defense of major cities, while each city is unique, all urban areas with a large population (e.g., over 500,000 people) share features. For example, each large city:

- Supports the lives of hundreds of thousands of people (water, sewer, power, transportation, energy),
- Operates a major airport,
- Has major highways and railways that either run through the city or next to it,

---

[65] Health and Human Services Press Office, "News Release: Guidelines for Bioterrorism Funding Announced," May 9, 2003. Available at <http://www hhs.gov/news/press/2003pres/20030509.html>.
[66] Health and Human Services Press Office, "News Release: HHS Announces $1.1 billion in Funding to States for Bioterrorism Preparedness," January 31, 2002. Available at <http://www.hhs.gov/news/press/2002pres/20020131b.html>.

- Runs schools,

- Houses government buildings, a jail or prison, at least one large academic institution, and a stadium and/or other sports arena,

- Depends on centers of commerce and maintains open access to them,

- Supports local media (e.g., radio stations and at least one local TV station),

- Offers services in major hospitals and emergency treatment centers, and

- Hosts major events (e.g., state fairs, concerts, holiday parades, New Year or 4th of July celebrations, races, etc.), some regularly or annually and some on a one-time basis.

In addition, some cities themselves share common features (e.g., all coastal cities have ports and boat access, and all cities that cross a major river have at least one major bridge).

Given these similarities, it is remarkable that states and cities currently develop their emergency plans almost completely independently. This generally means that they do not learn from the efforts and experiences of others, which ultimately may translate into wasted resources. In addition, development of processes and training materials do not benefit from sharing, and much the same is true of smaller cities and rural areas. Further, because of the focus on major metropolitan areas, rural areas are a continued weak link.[67] Regional response planning is minimal at best.

Recommendations

- *Current efforts to train responders should explicitly consider the level of training needed at each level of government as well as integration of new procedures. The efforts should also explore the opportunities to share some of the analytical work (i.e., modeling) where generic models would provide useful tools. Better coordination between military and civilian responders and evaluation and peer review could lead to significant improvement of plans and cross-fertilization of good ideas. Focus on anthrax as a single agent for discussion is likely to lead to the identification of key issues that existing generic plans might miss.*

- *HHS should pioneer an effort to develop national preparedness standards—including local, state, and regional coordination requirements—and should tie Federal grants to meeting these requirements.*

---

[67] Gursky, 1-4.

Research portfolio

The NIH, CDC and DOD all maintain active research efforts related to anthrax, but they lack overall coordination and largely focus on basic science and on promoting local public health or military preparedness. These research efforts continue to make progress that furthers our understanding of anthrax and our abilities to deal with it. However, the research to date includes relatively little focus on quantitative evaluation of the risks, costs and benefits of different actions. Consequently, national decisionmakers continue to lack the policy tools they need to allocate scarce resources properly.

Recommendations

- *The government needs to develop the analytical methods required to evaluate its investments in basic science and technology development. These methods should include mechanisms for improving local, state and Federal coordination in planning and response, as well as ways to measure the benefit that may be achieved from such efforts. While significant levels of support exist, continued support may depend on demonstration of real benefit and reasonable trade-offs between costs and benefits. Demand for these analyses may also significantly advance efforts to explore dual-use opportunities and to encourage better information sharing and coordination between military and non-military researchers and policymakers.*
- *The National Academy of Sciences should be asked to evaluate the efforts made thus far—in government, industry and academia—to address issues of bioterrorism in general and anthrax in particular. The Academy should be asked to propose a quantitative evaluation process to help prioritize expenditures directed to bioterrorism and public health preparedness.*

EXAMPLE OF AN ANTHRAX RESEARCH INVESTMENT THAT RECENTLY PAID OFF[1]

FDA approved a new test, produced by Immunetics, Inc., and funded by the CDC, to rapidly and easily determine whether patients have been infected with anthrax. Called the Anthrax Quick ELISA test, the test takes less than one hour. It detects an immune response to a protein produced by the infecting anthrax bacteria. Availability of this test means that state and private laboratories can test to rule out anthrax instead of needing to send every sample to either the CDC or the U.S. Army (USAMRIID).

[1] CDC Media Relations, "CDC Collaboration Yields New Test for Anthrax," June 7, 2004. Available at <http://www.cdc.gov/od/oc/media/pressrel/r040607.htm>.

Information Management

With the staggering amount of information being generated, organization emerges as a critical issue. Inventories of resources and capabilities are important tools in coordination. The inventory of anthrax laboratories represents an important improvement in information management; other tools should include flowcharts with decision points and identified information needs that can aid in coordination of responding to an event.[68]

Recommendations

- *For anthrax, much of the needed information management within the CDC and other government agencies exists. Coordination with other stakeholders now is a major priority. As information increasingly becomes available on the Internet, stakeholders will need tools to sort through and find the high-quality information that they need.*
- *Professional medical societies should develop clinical protocols for early detection, response, treatment, and reporting of biological warfare cases in coordination with CDC requirements.*

Policy development

A number of policy issues arise in the context of managing the risks from biological weapons. The ability to foresee many of these risks leads to opportunities to address (some of) them and engage in debates without the pressure of a crisis. The legal issues span a huge range. They include clearly identifying roles and responsibilities so that all stakeholders know who has which authorities and responsibilities to make decisions and act, and the limits on these. Key issues of concern include quarantine policies and restrictions on freedom, including limiting access to private property and compensation for use of private property.

With all of the demands for coordination and information sharing, issues inevitably arise regarding individual privacy and the need to protect the public. Federal laws, such as the Health Insurance Portability and Accountability Act of 1996 (HIPAA), create an important context for some individuals. Protecting emergency responders and

---

[68] CDC, "A National Public Health Strategy for Terrorism Preparedness and Response, 2003-2008" (March 2004): 4. Available at <http://www.mipt.org/pdf/National-Public-Health-Strategy-Terrorism-Preparedness-Response-2003-2008.pdf>.

cleanup contractors from liability may also be critical to creating the right incentives for them to help. For example, if doctors are asked to help in the response, but lack adequate liability protection, patient care may be degraded. With respect to vaccines, ethical issues arise if vaccination is not voluntary. Ethical issues may also arise if authorities must ration scarce resources and deny treatment to some. Finally, with respect to our own national behavior, we must consider the issues that arise from any actions taken that appear not in good faith with respect to international agreements, even if we know other parties are in violation.

Recommendations

- *For anthrax, policymakers should decide which legal, ethical, and other policy issues would benefit from proactive efforts to develop policies before a crisis, and which issues to leave until the time of a crisis.*
- *The appropriate authority should charter a commission to identify the ethical, legal, privacy and civil liberties issues associated with a national response—not just a Federal response—to a health crisis. This evaluation should identify policy gaps and make specific recommendations for any legislative and/or regulatory solutions.*

Integration

In surveying the decisions related to risk management for anthrax, we see a complicated, but manageable picture. If we think of all of the decisions and bits of information as the pieces of a puzzle, then one perspective is that the decision tree provides the helpful box-top image of how they all fit together.

The major theme that emerges is one of using decision analysis to make better choices, and doing a better job characterizing the risks to promote cost-effective efforts that address the correct priorities. Decisionmakers need to focus on concrete outcomes that can be measured or modeled, including mortality and morbidity prevented and dollars and time spent or saved. Recognizing that uncertainty exists due to our lack of knowledge, we must move toward better characterization of critical uncertainties and use the concepts of value-of-information analysis to evaluate our investments of resources in opportunities to reduce these uncertainties.[69] In the context of pubic health today, the risk of widespread anthrax remains small, while every year common infectious diseases like flu take a significant toll. Efforts to improve preparedness for a bioterrorism event must recognize that some stakeholders may have greater priorities, and this should lead to discussions about the set of expectations and objectives that drive choices.

Summary

Many efforts are underway in the Federal Government to prepare for a bioterrorism attack. Some argue that, three years after the anthrax attack, we are less prepared than ever.[70] This study does not include a comprehensive review of the government's preparedness efforts and does not comment on the current degree of biodefense readiness—or lack thereof. However, anecdotal reporting from colleagues and personal experiences suggest that readiness could be significantly improved by the adoption of standardized approaches.

The risk management approach recommended in this paper offers a proven, effective way of "organizing for combat" in the public health arena. Introducing this approach piecemeal into every organization concerned with biodefense is unrealistic. Moreover, there is no single government official who speaks with final authority on the topic of biodefense and who could direct the adoption of such a framework. Introducing it into the interagency process through "wargames," drills, practices, exercises, etc., seems the most efficient way to disseminate the methodology.

Specifically, on the topic of anthrax, this study makes no bold or sweeping recommendations. Rather, it argues that anthrax must be viewed in the larger framework of possible and probable infectious diseases to which the human race is exposed—both naturally and as a result of deliberate terrorist activity. However, given our recent experience with anthrax, the specific decision trees for anthrax (see pp. 32-33) are offered

---

[69] Fumie Yokota and Kimberly M. Thompson, "Value of Information (VOI) analysis in Environmental Health Risk Management (EHRM)," *Risk Analysis* 24, no. 3 (2004): 287-298.

[70] John Mintz and Joby Warrick, "U.S. Unprepared Despite Progress, Experts Say," *Washington Post*, November 8, 2004. Available at <http://www.washingtonpost.com/wp-dyn/articles/A32738-2004Nov7.html>.

as an analytical tool to aid future policy decisions. Indeed, the lessons learned from the 2001 attack should facilitate the use of these trees.

Of all the topics covered in this decision tree approach, media preparation appears to be the one with the potential for greatest and most immediate return on investment. There is a clearly identified need for public education messages, using nationally recognized and respected spokespersons, to address topics related to bioterrorism. Some of the messages would be generic and applicable to any disease outbreak—natural or man-made—while others would be specific to a given organism/disease. The strongest recommendation of this study is that decisionmakers at HHS, in conjunction with its subordinate organization the CDC, begin the preparation of such messages—but only with close cooperation of clinicians involved in healthcare delivery in the private sector (see p. 44).[71] The effort to curb the overuse of antibiotics is an excellent example of how public health education campaigns can be very effective in changing behavior.[72]

---

[71] Elin A. Gursky, Thomas V. Inglesby and Tara O'Toole, "Anthrax 2001: Observations on the Medical and Public Health Response," *Biosecurity and Bioterrorism: Biodefense Strategy, Practice, and Science* 1, no. 2 (November 2, 2003): 100. Available at <http://www.homelandsecurity.org/bulletin/Anthrax202001.pdf>.
[72] Background on antibiotic resistance available at <http://www.cdc.gov/drugresistance/community/>.

Appendix A – NDU Anthrax Conference

In May 2004, Dr. Anna Johnson-Winegar, former Deputy to the Assistant Secretary of Defense for Chemical and Biological Defense, hosted a senior level seminar at the National Defense University on the topic of anthrax. The seminar, sponsored by the Center for Technology and National Security Policy and entitled "From A to X: An End-to-End Review of Anthrax," reviewed various topics related to anthrax. (This study grew out of one of the presentations at the seminar.)

Dr. Johnson-Winegar's summary follows as Appendix B. A list of the presenters is provided below. All of their presentations are included on the CD inserted in the inside back cover of this report. The CD also contains a full list of participants and their organizations to aid continued discussion and coordination among professionals in this field.

CTNSP/NDU Senior Level Seminar Series
Anthrax Workshop
National Defense University
George C. Marshall Hall, Room 155

Wednesday, 12 May, 2004

0815	**Registration Available for Workshop Attendees (Continental Breakfast Served)**		
0900	Welcome / Workshop Introduction	Dr. Anna Johnson-Winegar	
0915	*Setting the Stage* "Biological Warfare: Where We've Been; Where We Need to Be"	Col Donald Thompson, MD, MPH&TM (U.S. Northern Command)	
1000	**Break**		
1015	**Panel Discussion – DOD and CDC Perspectives**	DOD Medical Policy	COL Terry Rauch, Health Affairs
		AVIP Program	COL Steve Jones, MVA Director
		Legal Issues and Procedures	John Casciotti, Office of General Counsel
		Civilian Welfare	Dr. Nina Marano, Center for Disease Control & Prevention
		Q&A for Panel	
1215	**Catered Lunch Courtesy of CTNSP**		
1245	**Luncheon Speaker**	Ms. Judith Miller, *New York Times*	
1330	**Risk Management and Integrated Decisionmaking**		Dr. Kim Thompson, Harvard School of Public Health
		Q&A	
1415	**Break**		
1430	**Panel Discussion – Policy Concerns for the Future**	Department of Homeland Security	Dr. Carol Linden, Science Based Threat Assessment and Response
		Impact on the Mail	John Bridges, U.S. Postal Service
		First Responders	James Rohan, U.S. Capitol Police
		Anthrax & Public Policy	Dr. Stephen Prior, NSHPC, Potomac Institute for Policy Studies
		Q&A for Panel	
1630	**Wrap-up**		

Appendix B – Johnson-Winegar Review

From A to X: An End-to-End Review of Anthrax
Anna Johnson-Winegar, Ph.D.

<u>Introduction</u>

Anthrax has been studied for centuries—yet we still have much to learn. In the early days of microbiology, Robert Koch, Louis Pasteur and others made observations on the physical characteristics of *Bacillus anthracis*, the causative organism. Anthrax was known primarily as a disease of cattle, sheep, and other types of livestock, but it also infects others species, including monkeys and humans. In 1876, Koch published his pivotal work on "Koch's Postulates," which provided the basic methodology used by scientists and clinicians to link a specific bacterium with a resultant specific disease. Not only was anthrax the case study for Koch, but it was also the first bacterium for which a vaccine was developed. In 1881, Louis Pasteur created the first vaccine for anthrax. Thus, mankind has been worried about the effects of anthrax for well over a century.

Anthrax infection in humans manifests itself in three ways: cutaneous, gastrointestinal, and inhalational (respiratory). These forms of the disease generally indicate the route of infection for the exposed individual. The cutaneous form of the disease is the most common, accounting for about 95 percent of all cases, and is also the most easily treated. The gastrointestinal form is more severe, and results from ingestion of contaminated meat (this was the original explanation given by the Soviets for the deaths following the Sverdlosk accident in 1979). The inhalational form of anthrax is by far the most deadly, approaching 100 percent fatality if untreated.

One unique aspect of the anthrax bacteria is the hardiness of the spore stage, and there have been reports of spores being viable after many decades in the soil. This is but one of the attributes that make anthrax the probable favorite of all the potential biological warfare agents. Other characteristics of an organism that make it a good choice for a BW agent are the following: ease of preparation; resistance to ultra-violet light; stability in various types of weapons systems; a relatively short incubation period; ability to be dispersed as a liquid or dry powder; and a small amount required for a mass casualty effect. Numerous estimates have been developed that indicate a few kilograms of anthrax, if delivered efficiently, could cause casualties in excess of one million individuals (assuming ideal weather conditions). Therefore, it is easy to see why anthrax has been the number one choice of a BW agent for both terrorists and organized nations or state sponsored groups. If, indeed, anthrax has been on the top of everyone's list for decades, why don't we have the problem solved? The answers are complex—they encompass both the scientific and political worlds.

The DOD Response to the Threat of Anthrax

Although anthrax had been at or near the top of the validated threat list since its inception, the DOD had not been aggressive in developing a posture to protect the military forces. Indeed, the Persian Gulf War of 1990-1991 was a strong wake-up call for the DOD. Never before had the senior leadership addressed the possibility of biological warfare so seriously. Numerous briefings were held on a regular basis for the Joint Chiefs of Staff (JCS), and when intelligence assessments indicated a strong possibility that Saddam Hussein possessed biological weapons (including anthrax), the DOD developed a comprehensive plan to afford maximum protection. Elements of the plan included deployment of rather rudimentary sensor devices to provide early warning and detection; individual protection (masks and protective overgarments); collective protection and shelters; and the initiation of immunizations in theater.

The decision to immunize military forces against anthrax was not an easy one. Again, numerous briefings were held with intelligence analysts, policy staff, military operators, and acquisition officials. Although military forces are routinely vaccinated for many endemic diseases, large-scale use of a vaccine against a biological warfare agent had never been conducted. The anthrax vaccine was developed under contract to the U.S. Army, and was subsequently produced and licensed in 1970 by the Michigan Department of Public Health (MDPH). Prior to 1990, the MDPH had sold a few hundred doses of vaccine per year, primarily to at-risk laboratory workers, veterinarians, and some workers in the wool sorter and animal hides industry who might come in contact with contaminated materials. One of the logistical issues surrounding this vaccine was the need for six shots given over a period of eighteen months to provide full immunity as described in the license. The MDPH facility increased production to the best of its ability; however, the production process is time consuming and constrained by biological issues (e.g. a 30 day potency test in guinea pigs). Since the amount of vaccine that had been stockpiled was minimal, the DOD began its immunization program with less than a full supply of vaccine for total forces protection. Due to the short duration of the conflict, most military members actually received only one or two doses of vaccine in theater. The commitment on the part of the DOD to complete the immunization series for those individuals became a difficult issue to resolve with the Food and Drug Administration.

The Food and Drug Administration (FDA) is the regulatory agency with purview over vaccines. It maintains responsibility for approving the release of each lot of manufactured product and for following and adverse reactions to the immunization. The DOD worked closely with the FDA to describe the specifics of the situation and clarify the intent to provide the full immunization series to individuals who had started the series. Adequate supply of vaccine was the primary factor in the inability to complete the program that had been initiated. Inadequate supply also precluded the DOD from providing anthrax vaccine to Allies and coalition partners. Following the Gulf War, the DOD continued to evaluate the pros and cons of large scale immunization. Following recommendations from senior staff, including the Assistant to the Secretary of Defense (Health Affairs), and others, Secretary of Defense Cohen announced a decision for immunization of the total force in 1998. Some of the rationale for such a decision was the rotation of troops;

the long period required to achieve full immunity; the desire to conduct immunizations prior to any deployment; and the necessity to maintain optimal readiness. This decision included a mandate for an external review of the program by subject matter experts.

During the implementation of the anthrax vaccination policy, numerous legal issues arose. Some challenged the lawful order and the right of the individual to consent. Others addressed the issue of whether the vaccine should be considered investigational since the license did not specifically state that the vaccine protected against an aerosol challenge with anthrax spores. Several individuals developed medical complications following their vaccination and attributed their disability to a direct cause of the vaccine. These claims were not upheld in court and the Presidential Advisory Commission on Gulf War Illness (GWI) indicated they found no direct linkage of anthrax vaccination to the cause of GWI. Further study by the Institute of Medicine found the anthrax vaccine to be safe and effective. Later in 2000 and 2002, the policy was amended to indicate vaccination only of those forces deployed early to high threat areas.

The DOD has become more aware of the need to address total force protection and has instituted more proactive measures to evaluate individual health issues prior to deployment, and immediately upon return from deployments. Although not a perfect solution, these steps toward understanding the consequences to one's health (both immediate and delayed) will go a long way in allaying concerns.

Anthrax Vaccination for the Civilian Sector

Although the DOD's anthrax immunization policy has been underway for several years, the possibility of a similar policy for civilians has not been seriously considered. Following the anthrax letters in the fall of 2001, the Department of Health and Human Services (HHS) (led by the Centers for Disease Control and Prevention) advised use of antibiotics following potential exposure. Since there were limited data available about the effectiveness of vaccine post exposure, there was not wide spread use of vaccine, even in those workers on Capitol Hill who had most likely been exposed to anthrax spores. Today, the HHS has decided to procure ample quantities of anthrax vaccine (approximately 25 million doses) for the Strategic National Stockpile with the intent to administer vaccine plus antibiotics post exposure. Some of the initial material in the stockpile is the currently licensed vaccine, produced by BioPort (successor to MDPH); however, the long-term strategy is to use a new recombinant vaccine, based on the protective antigen (PA) factor of the anthrax toxin. First, however, FDA must license any new vaccine (recombinant or other). In the absence of the ability to conduct clinical efficacy trials, it will be necessary to establish clear criteria for surrogate markers in animal studies. Second, since post-exposure use of vaccine to treat anthrax is not currently licensed by the FDA, these types of regimens would have to comply with the provisions of the Investigational New Drug (IND) regulations. While this may be feasible in a small controlled scenario, it is most likely not possible in a large scale outbreak, and the DOD has found immense difficulty in complying with these provisions in a combat setting. Full approval and licensing of anthrax vaccine for post exposure use will require specific clinical studies for safety and immunogenicity. These studies must compare use of extended use of antibiotics alone and in combination with vaccine. National security issues regarding the strategic national stockpile of medical countermeasures (to include anthrax vaccine and antibiotics) are being transferred from the HHS to the Department of Homeland Security.

Understanding the Threat

One of the primary tasks facing the National Biosecurity Analysis and Countermeasures Center (NBACC) is the thorough evaluation of the threat from anthrax and other biological warfare agents. Scientists will work in conjunction with intelligence analysts to perform appropriate modeling and simulation studies, in conjunction with laboratory tests to learn more about important criteria. One of the major unanswered questions posed to date includes a reaffirmation of the lethal dose of spores for humans. The currently espoused range of 8000-10000 has come into question following the delivery of the anthrax letters through the postal system. It appears likely that some of the victims may have received a much smaller dose than that previously thought to be fatal. While it is difficult to ascertain the dose delivered to a specific individual, it becomes critical in understating the pathogenesis of the disease, the need for more sensitive detection and identification systems and appropriate medical countermeasures. These types of studies will help characterize the vulnerability of the population.

Another primary thrust for the NBACC is the forensic analysis required following any type of use of a biological warfare agent. Various types of analyses and signatures attributed to select strains of *Bacillus anthracis* and other pathogens need to be analyzed and developed into data bases appropriate for data mining. Knowledge management is a key component of pulling together the necessary bits and pieces of information that need to be connected for a comprehensive study.

Another aspect of analyzing the threat is the current program in biosurveillance and monitoring. The DHS has established BioWatch in selected urban areas and will continue to collect valuable data related to background levels of organisms. In conjunction, there is a comprehensive effort underway to monitor increases in disease outbreaks through local health clinics, private physicians' offices, school absences; and over-the-counter sales of specific medications (including those for treating upper respiratory infections, diarrhea, and headaches). While it is too early to critically evaluate the potential usefulness of these surveillance systems, efforts such as these to bolster the public health infrastructure and methods to improve faster communication of data represent essential elements of an overall strategy.

From Hypothesis to Reality

While the specter of biological warfare had been looming for many years, the actual delivery of anthrax spores through the postal service made the nation into believers. From the initial case in Florida, through the additional cases in the Brentwood postal facility in Washington, D.C. on to the letters delivered to Capitol Hill, the attention of the American public was captured.

Two aspects dominated the attention of the postal service and the capitol police in dealing with the anthrax letters. The first dealt with obtaining confirmatory evidence of the identification of the material. Initial assays were performed using the hand-held "tickets," a system based on immunologic reagents. Confirmatory tests were provided by laboratory analysis, using immunoassays, polymerase chain reaction, growth of samples on cultures plates and visual examinations. The need for analysis of literally tens of thousands of samples clearly highlighted the sparse capacity of the laboratory infrastructure available. Presumptive testing proved to be sufficient for most of the decisionmaking. The large number of both environmental samples (i.e. swipes from furniture, etc), and clinical samples for diagnosis (e.g. nasal swabs) inundated the available labs in the area. The need for more laboratory testing capacity through the public health system (including state health laboratories) is being addressed through additional funding provided to HHS and administered through state and community grants. Under the auspices of the CDC, a tiered structure has been put in place allowing local labs to do initial screening, with reference laboratories scattered throughout the country that are able to do more sophisticated tests and to handle BL-4 pathogens.

Decontamination of infected areas at the postal facility and the Senate office building proved to be a large challenge. As already noted, anthrax spores are extremely hardy and resistant to many types of decontaminating materials that are appropriate for most biological agents. Eventually the use of chlorine dioxide, under controlled temperature and humidity, proved effective for the Hart Senate office building. More extensive testing was required to assure decontamination of the larger Brentwood postal facility. These efforts were extraordinarily expensive and consumed vast resources. Simultaneously, testing was done by DOD laboratories to validate the process of decontaminating the mail by irradiation. Again, the basic data available and knowledge of the technical complexity was insufficient. Although the DOD had been investing in decontamination of chemical and biological agents for decades, their focus had been on military vehicles and large open areas. No one had really considered appropriate methods of decontamination that would be effective against anthrax spores, yet safe for use on office furnishings (carpet, drapes, computers, etc). Today the Environmental Protection Agency is the lead Federal agency charged with the responsibility for decontamination, and research continues for a better decontaminant. Requirements vary, but desirable attributes include developing a decontaminant that is environmentally friendly; safe to use on humans; safe for use on sensitive electronic equipment; effective against a broad range of chemical, biological, and radiological agents; requires little or no water; is easy to disperse; works in a short period of time; and is relatively inexpensive.

<u>The Way Ahead</u>

Unfortunately, today, we still have many unanswered questions about these incidents involving the use of anthrax spores. We do not know who was responsible. We do not know the source of the anthrax spores (i.e. were they produced, stolen, or purchased?). We do not know the motive for the attacks. And, therefore, we are unable to make intelligent assessments about the likelihood of similar attacks in the future. Clearly there is a need for additional intelligence assessment and further technical understanding of the use of biological organisms as biological warfare agents—either by terrorists or by military adversaries.

Knowledge and communication have emerged as two primary focal areas for improvement. It is incredibly important for the public to be kept informed in order to lessen the possibility of panic. It is critical for the responsible government officials to provide accurate information in a timely fashion. We can diffuse terror and panic by increasing situational awareness. In the absence of solid information, confusion reigns.

One approach toward sharing knowledge and improving communication is through exercises- both table-top and actual field training. The first responders to an incident in the civilian population will be the local fire, police, emergency medical technicians, and others. This community, as a whole, is in need of substantially higher funding for education, training, and equipment. The Federal, state, and local entities must work together to share knowledge and improve responsiveness.

Learning how to manage the risk associated with the use of anthrax (or any other biological agent used by a terrorist) is among the highest priorities. Many decisions must be made in a short period of time by numerous individuals or organizations that may have been previously unconnected. Development of a matrix that covers broad topics is a logical first step. Some examples that would be considered in this matrix include the following:

Many decisions....

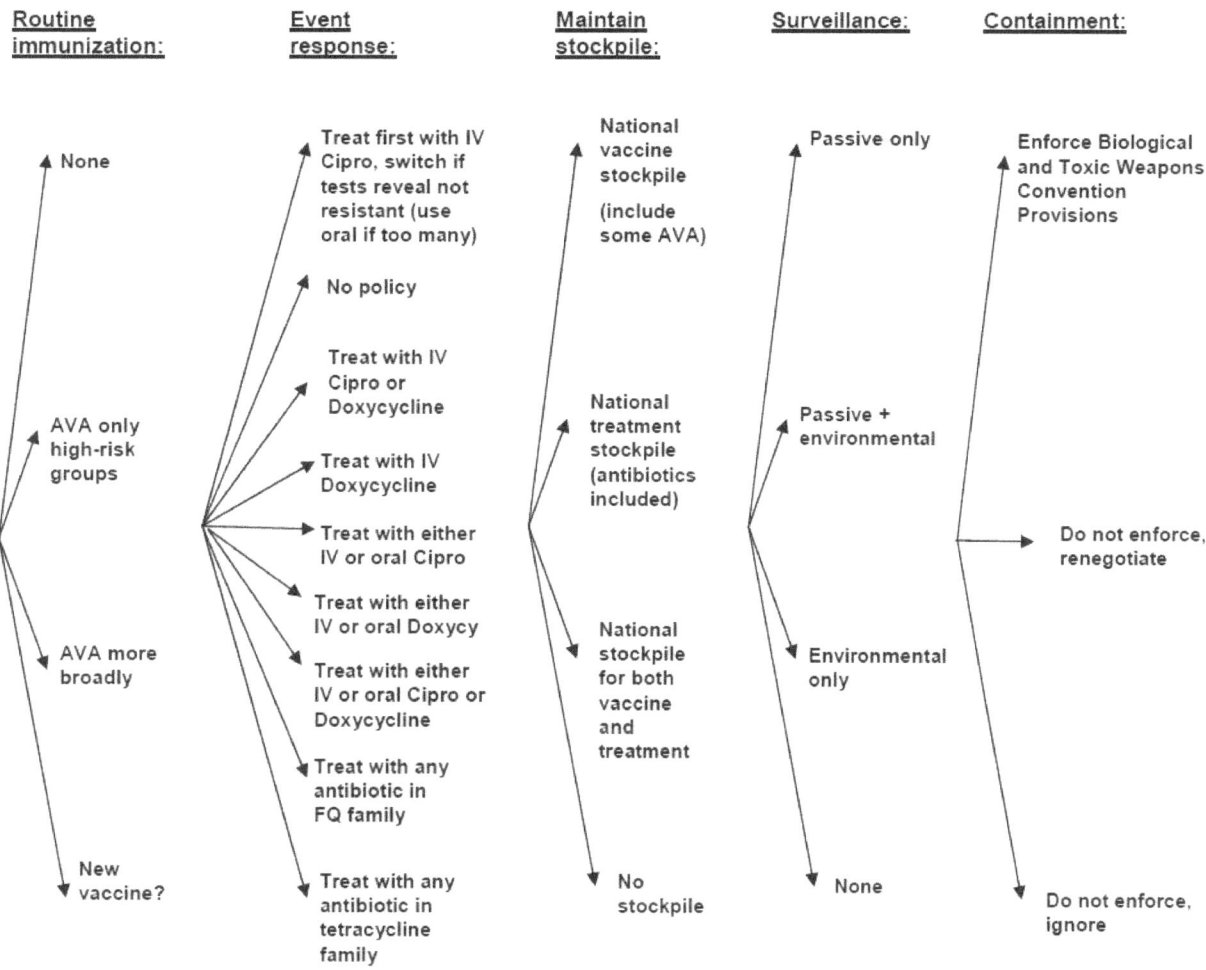

Routine immunization:

- None
- AVA only high-risk groups
- AVA more broadly
- New vaccine?

Event response:

- Treat first with IV Cipro, switch if tests reveal not resistant (use oral if too many)
- No policy
- Treat with IV Cipro or Doxycycline
- Treat with IV Doxycycline
- Treat with either IV or oral Cipro
- Treat with either IV or oral Doxycy
- Treat with either IV or oral Cipro or Doxycycline
- Treat with any antibiotic in FQ family
- Treat with any antibiotic in tetracycline family

Maintain stockpile:

- National vaccine stockpile (include some AVA)
- National treatment stockpile (antibiotics included)
- National stockpile for both vaccine and treatment
- No stockpile

Surveillance:

- Passive only
- Passive + environmental
- Environmental only
- None

Containment:

- Enforce Biological and Toxic Weapons Convention Provisions
- Do not enforce, renegotiate
- Do not enforce, ignore

59

Many decisions....

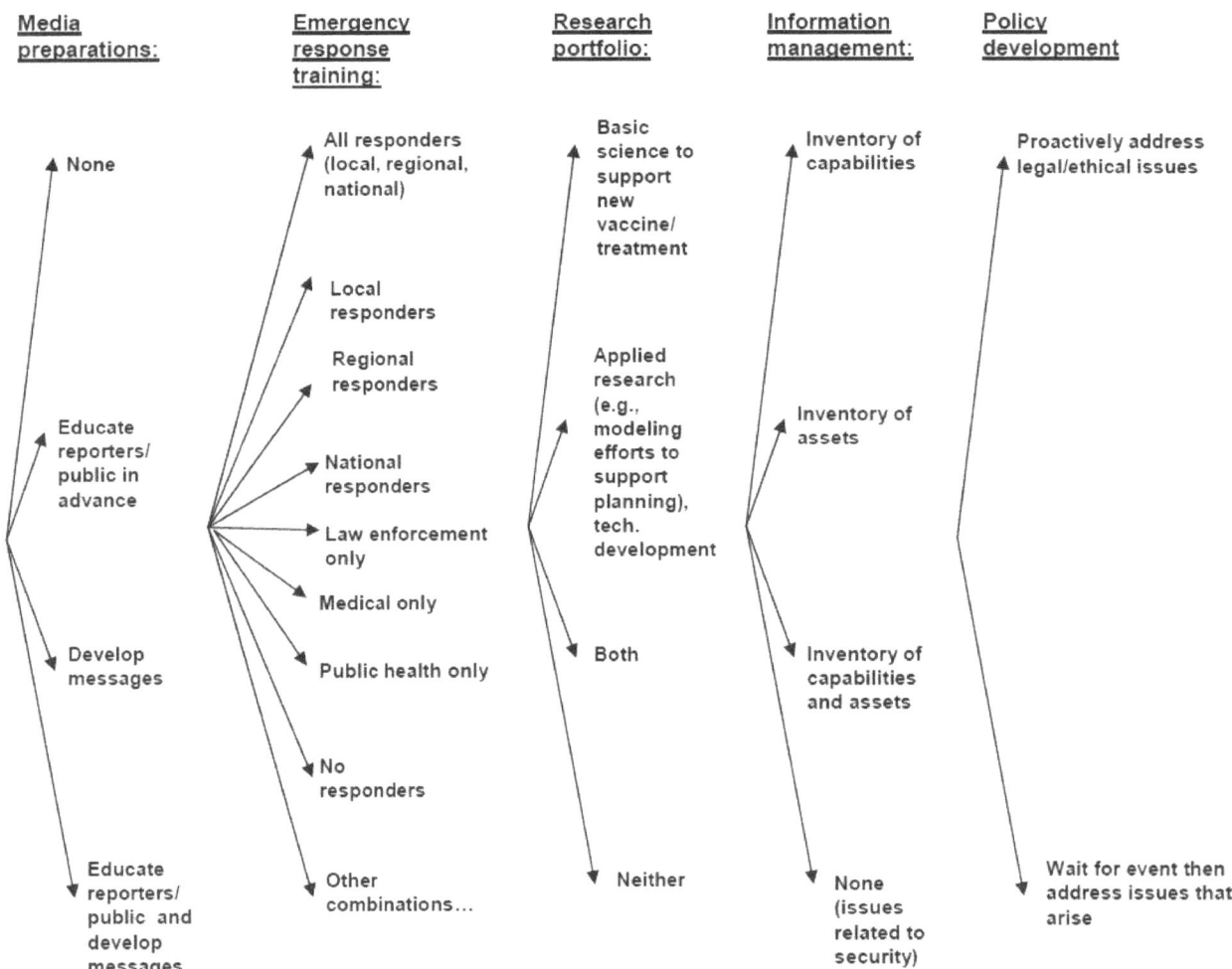

Media preparations:
- None
- Educate reporters/ public in advance
- Develop messages
- Educate reporters/ public and develop messages

Emergency response training:
- All responders (local, regional, national)
- Local responders
- Regional responders
- National responders
- Law enforcement only
- Medical only
- Public health only
- No responders
- Other combinations...

Research portfolio:
- Basic science to support new vaccine/ treatment
- Applied research (e.g., modeling efforts to support planning), tech. development
- Both
- Neither

Information management:
- Inventory of capabilities
- Inventory of assets
- Inventory of capabilities and assets
- None (issues related to security)

Policy development
- Proactively address legal/ethical issues
- Wait for event then address issues that arise

In order to develop a cohesive national policy and strategy, it is incumbent for the various Departments of the Administration to work together for each has been assigned various responsibilities in the National Response Plan and outlined in other Presidential Decision Directives. In addition, the state and local agencies and departments must be incorporated into a plan.

Anthrax has long been recognized as a biological threat (in fact many experts agree it remains the number one threat from both terrorists and military adversaries). While we have learned a great deal over the past few years (as evidenced by the increased number of publications in the scientific literature and the greater understanding of the lay public), there is more to be done. Some of the detailed research needed to understand more about this pathogen will take years to complete; other actions can be completed in the short term.

The concept of eradication of disease has been presented to the public in terms of naturally occurring smallpox. In addition, there is currently an effort underway to eradicate polio from the earth. These campaigns can be successful since there are no other natural reservoirs for these viruses; however, anthrax remains a zoonotic disease and it would be extremely difficult to remove all natural sources of this bacterium from the earth, without even giving consideration to stockpiles of spores that may be held by potential enemies or terrorists. Therefore, if one acknowledges that some form of anthrax will always be present, one must concentrate on managing the situation rather than attempting to totally avoid it.

Improved education of the public and senior government officials is critical. A better understanding of the possible consequences will eliminate panic, and will facilitate emergency planning. Training exercises are essential for all members of the first responder communities that may be faced with an anthrax situation. Increased research efforts must be continued to improve the sensitivity and specificity of detectors. A valid concept of operations must be implemented following a positive detection alarm. Better medical countermeasures are needed to provide protection in advance (improved vaccine) and for post exposure therapy (possibly anti-toxins or generic drugs that interfere with toxin binding) as well as definitive clinical studies on the value of antibiotics and vaccine post exposure. Finally, decontamination remains a key issue when considering anthrax because of the durability of the spores. Comprehensive programs must be developed; programs must be prioritized; support and funding must be sustained; and intelligence must be constantly evaluated. Anthrax has been with us for centuries. It is not an easy task to "take it off the table" as a choice of biological weapons for either terrorists or military adversaries. It will require commitment and sustained interest from all involved.

www.ingramcontent.com/pod-product-compliance
Lightning Source LLC
Chambersburg PA
CBHW081606170526
45166CB00009B/2856